Cavers, Caves, and Caving

Bruce Sloane received his A.B. degree from Dartmouth College and his M.S. degree from Montana State University. He is the son of Howard N. Sloane, who with Charles E. Mohr edited *Celebrated American Caves*, published by Rutgers University Press.

Cavers, Caves, and Caving

Edited by Bruce Sloane

RUTGERS UNIVERSITY PRESS
New Brunswick New Jersey

Grateful appreciation is expressed to the following for their courtesy in permitting the reprinting of copyrighted material:

To Russell H. Gurnee, president of The Explorers Club, and *Explorers Journal* for "Exploration of the Tanama" by Russell H. Gurnee, copyright © 1971. To George F. Jackson for passages from *The Story of Wyandotte Cave*, copyright © 1975. To Harold Meloy for passages from *Mummies of Mammoth Cave*, copyright © 1968, and for "Stephen Bishop: The Man and the Legend," copyright © 1974. To *Natural History* magazine for "Refugees of the Ice Age" by Thomas C. Barr, Jr., reprinted with permission from *Natural History* magazine, May, 1974, copyright © The American Museum of Natural History, 1975. To *Natural History* magazine for "Sloth Droppings" by Paul S. Martin, reprinted with permission from *Natural History* magazine, August-September, 1975, copyright © The American Museum of Natural History, 1975.

Library of Congress Cataloging in Publication Data
Main entry under title:

Cavers, caves, and caving.

Bibliography: p.
Includes index.
1. Caves—North America—Addresses, essays, lectures. 2. Speleology—Addresses, essays, lectures. I. Sloane, Bruce, 1935-

GB603.C38 917.3'09'44 76-55358
ISBN 0-8135-0835-5

First Printing

Manufactured in the United States of America

In memory of my father, Howard N. Sloane

Contents

Foreword

Beneath America's earth lies a remote, dark, silent wilderness known well by relatively few people—the world of caves. Ask the man in the street if he supports the preservation of caves and he is most likely to answer "Preservation of *what?*" If you pursue the matter you will probably find that he may have visited a few commercialized caves and found them interesting, but he would not want to confront the dark, the water and mud, and the tight passages of the true cave environment. But for a small percentage of the population caves are a joy, if not an obsession, that they prefer beyond open forests, mountains, rivers, and, indeed, daylight. This book is by and about such people—the cavers and speleologists—and the environment they explore and study.

Several hundred books have been written about American caves since 1882 when the Reverend Horace C. Hovey published *Celebrated American Caverns*. The books range from adventure stories for children, through how-to-do-it books, to scientific treatises. Many more stories and articles circulate among cavers and speleologists, but little of this is seen by the general reading public. It was not until 1955 that the first anthology of cave stories, *Celebrated American Caves*, was published by the Rutgers University Press, edited by Charles E. Mohr and Howard N. Sloane—Bruce Sloane's father. That this anthology continues, or rather starts, a "family" tradition, is as fascinating a story as any contained herein.

Those interested in caves come from every walk in life. In the course of time they often come to specialize in some aspect of caves. Because caves are really environments rather than "things," the whole range of environmental interests is found among those who explore and study them. Archaeology and anthropology, biology, ecology, geology and geography, history, hydrology, mineralology, exploration and surveying, paleontology—these are a few of the disciplines in which cave studies are found. Traces of all of these will be found in these stories, but it is the insight and enthusiasm of the authors that transform disciplines into fascinating accounts of what is found, what happens, and what has happened in the voids beneath our feet.

Another concern that underlies most of what is written today about caves is the desire of those who have discovered their complex fascination to protect and preserve the underground environment for the enjoyment and education of future generations. Concern for our natural environments has been a strong individual motivation and political force in this decade. Cavers and speleologists alike have suddenly found their long-standing concerns for a fragile environment to be part of a larger movement. This has been a welcome and mutually beneficial development. It is no simple task to halt the depredations of vandals, the poisoning of our groundwater supplies by the dumping of wastes into cave systems, the cutting of the delicately balanced threads of life that are unique to caves, and the impingement of a growing technological society on this rare and easily damaged environment.

From mutual interest and for mutual support, persons interested in caves founded the National Speleological Society (NSS) in 1941. Today the NSS has nearly five thousand members in the United States. It provides communication among cavers and speleologists with any and all interests in caves, and acts upon their common desires to learn about, enjoy, and protect caves. It is consistent with these common bonds that all the authors in this anthology (including several past presidents), and the editor, are associated with the NSS. It has been my special privilege to have met and worked with many of them on shared goals.

All of these stories share three characteristics: a wealth of information, an enthusiastic but concerned attitude toward caves and the phenomena they contain, and a desire to share personal experiences and enjoyment. Look for these, and your own reading pleasure will be magnified. It is with that hope in mind that I introduce you to Bruce Sloane and his fellow authors.

Rane L. Curl
President, National Speleological Society
1970-1974

Editor's Note

More than twenty years have passed since the publication of the first general anthology on speleology, *Celebrated American Caves*, edited by Charles E. Mohr and Howard N. Sloane. Caving and speleology have forged ahead in the intervening years. The National Speleological Society has grown from a handful of hobbyists and a few dedicated speleologists to a respected scientific body (although the hobbyists remain the backbone, and often are the speleologists, too). Hopefully, there will always be new caves and new discoveries in old ones to explore, map, and describe, but the emphasis today is more on understanding and synthesis than on pure description. Although solution caves in limestone and related rocks continue to outnumber others by far, more and more attention is turning to caves developed in lava, glacial ice, talus, and other materials. Computers now help map caves, digesting the data of cave systems 200 or more miles long that are too complex for any one person to comprehend. Articles on all aspects of speleology appear with regularity in learned journals. State cave surveys have been published for scores of areas. Studies of cave fauna have yielded new and useful information on speciation, ecology, and physiology. Interest in the cultural and social aspects of speleology has added to the knowledge of the important roles caves have played in United States history and heritage.

Techniques of cave exploration have advanced greatly in methodology. Twenty years of experience in fielding expeditions by the Cave Research Foundation and other groups have helped to establish efficient and effective multi-team exploration methods. A greater understanding of speleogenesis, the process of cave formation, has spurred new finds, helping more people—as Bill Halliday, dean of bibliospeleologists, says—to "read the language" of the rocks. A revolution in vertical caving, with new tools, hardware, and methods, has made the ascent and descent of thousand-foot pits seem almost routine. There are even improved methods of lighting, as slowly, but perhaps inexorably, the familiar old carbide lamp flickers out, to be replaced by more efficient and brighter electric headlamps and gear.

Unfortunately, caves and their contents are an especially vulnerable part of the environment. Vandalism and pollution are obvious threats, but sometimes the mere intrusion of well-meaning cavers can disrupt the delicate ecological balance. Cavers spend more and more energy and time working on conservation and environmental problems. As long as the demands for recreational outlets continue to outstrip the still-growing population, threats to caves will not decrease. This is the major problem facing cavers and speleologists in the years ahead, and it should be attacked with all technical skills available, including knowledge from the sciences of human behavior and understanding.

A volume such as this depends on the efforts of many people. The contributors, whose book this really is, have with understanding put up with my demands, even those I myself felt were unreasonable. Russell and Jeanne Gurnee, who knew far better than I the hazards of bookmaking, have offered me support and stimulus from beginning to end. Rane Curl's many suggestions could fill several books by several people. Richard A. Watson thoughtfully read and criticized portions of the manuscript, and Roger W. Brucker took the time to list seventeen alternate titles. Bobby Crisman of Carlsbad Caverns National Park was most helpful in providing material and reviewing my chapter on Carlsbad.

Thanks are also due to Louis Simpson, Paul Rowley, Barbara Unger, Roioli Schweiker, the late Peter M. Hauer, Eugenio de Bellard-Pietri, Rob Stitt, and others. Finally, I want to thank my wife Joy, who has been patient and understanding more often than not, and who even goes into a cave with me once in a while.

February 1976 Bruce Sloane

Cavers, Caves, and Caving

The World Underground

BRUCE SLOANE

Caves are natural holes in the ground that people can enter. A cave can be a few feet long, or go on for miles and miles. It can be a flat horizontal passage, a vertical pit, or a mixture of both. Although some caves can be traversed by walking, in others it may be necessary to crawl on your hands and knees (or even your stomach), or to use a rope or ladder, a boat, or diving equipment. A few caves can be entered by car, railroad, or even helicopter. The majority of caves, including the largest ones known, were formed by the solution of limestone or other soluble rocks, but there are many extensive lava caves. Caves can also be formed by streams in glacial ice, by the piling up of boulders or talus, by the action of ocean waves along ancient or present-day coasts, and by other means.

Caves may be bare, but more often they are decorated with stalactites, stalagmites, and other speleothems (formations). There may be delicate little grottoes or huge rooms. Bats, fish, insects, and other animals may live in caves, and remains of extinct animals that once used them may be found. Some caves

---

A caver, rappelling into an unexplored pit, does not know if he will find a dead end or a virgin passage at the bottom. ***Hodag C. Carbide.***

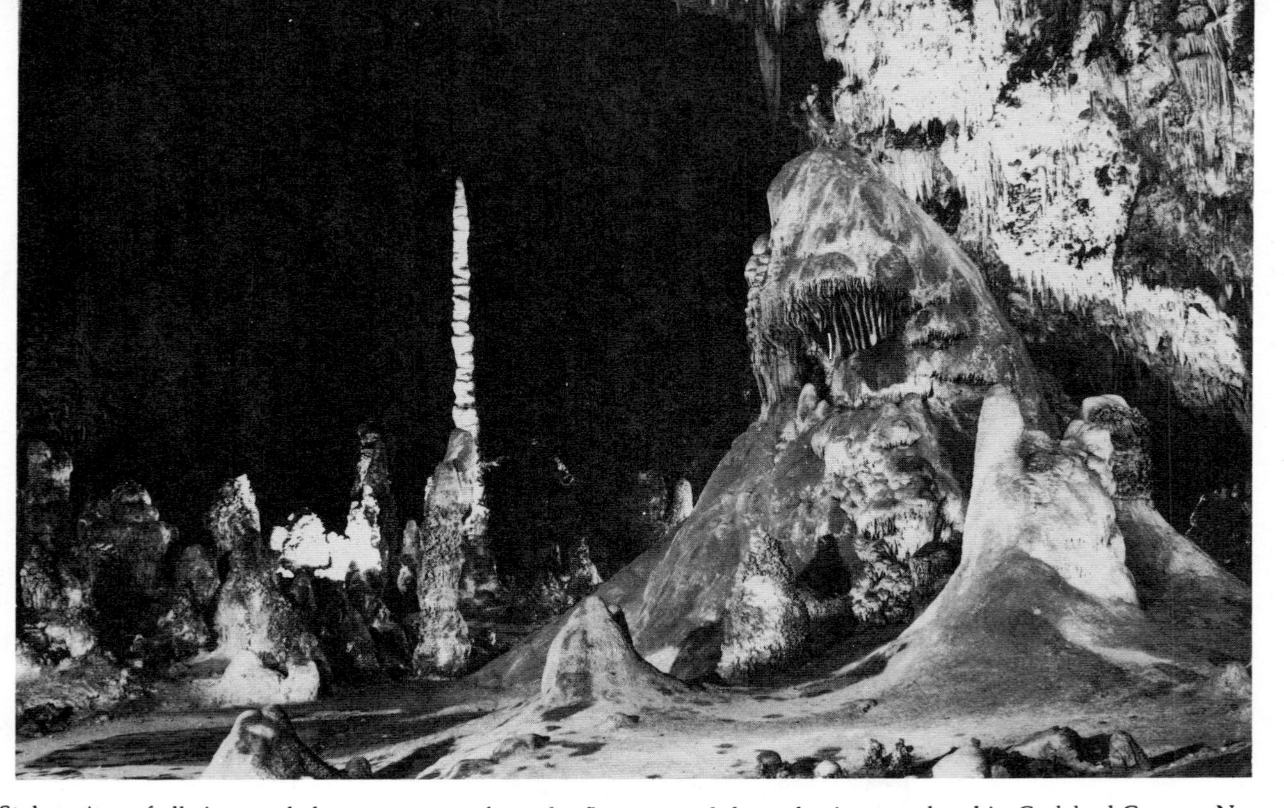

Stalagmites of all sizes and shapes grow up from the floor toward the stalactites overhead in Carlsbad Cavern, New Mexico. *National Park Service.*

have been visited by Indians, surveyed by George Washington, mined for guano for fertilizer or for saltpeter for gunpowder, served as storage for farmers' crops, or used as research laboratories. They may be open to the public and equipped with guides, lights, and walkways, or they may be undeveloped, requiring special equipment and techniques. Each cave is different from every other, and all of them are interesting.

Surprisingly, it is difficult to give a precise definition of a cave. Mines are not caves, since they are man-made, not natural. A mine may occasionally penetrate a cave, and many caves have been used as mines, but usually there is no problem telling the two apart. Purists insist a hole cannot be a cave if a person is unable to enter it for an appreciable distance. But if the entrance is a tight crawlway or squeezeway, a smaller, slimmer, or more supple caver may get through. The usual definition of a cave includes the qualifier that people can enter it.

This anthropomorphic orientation ignores several important features about caves. A cave is not an object; it is a space. Portions of a cave inaccessible to humans often are the home of many small animals, particularly invertebrates. Larger cave passages may be connected by smaller ones, impassable by humans, but serving as conduits for air, water, food, and organisms. A cave may not even have an entrance and still be important to the ecology and drainage of an area.

The scientific study of caves and related phenomena is called *speleology*, from the Greek word, *spelaion*, cave. A scientist who studies caves is a *speleologist*, and there are myriad "speleo-" word combinations whose meanings are clear enough from the rest of the word, despite the propensity of some to coin long definitions. An individual who studies the life in a lava cave, for instance, might be called a *volcanobiospeleologist*.

Someone who explores caves for the fun or sport of it is a *caver*, sometimes called a *spelunker*, and indulging in that spelean activity is called *caving*, or—occasionally—*spelunking*. Many cavers are also speleologists, but if a fledgling speleologist finds he is not also a caver he usually looks for another line of work.

Some caves are so well known that their names bring instant recognition: Mammoth Cave, Kentucky; Carlsbad Cavern, New

Flowstone covers most of the stalagmites in Lewis and Clark Cavern, Montana. *Montana Department of Fish and Game.*

Mexico; the Cavern of Luray, Virginia; and perhaps a handful more. These caves are exceptional and outstanding. But there are other caves—some unknown, some little known—that deserve wider recognition for their beauty, scientific interest, history, or associated tale of adventure or folklore. There are also fascinating stories of the people who explored and developed caves, of the animals that live there now or inhabited them in the past, of the environmental threats to caves and to their contents, and of the work of the speleologists who study them.

The stories in this volume have been gathered from many sources to provide a well-rounded view of American speleology today. Some were written specially for this work, while others have appeared before in a variety of publications.

There are plenty of adventure stories. The descent into the

Scuba divers carefully thread their way through the underwater maze of the Blue Holes of the Bahamas. *George J. Benjamin.*

400-foot pit of Alabama's Fern Cave eventually led to more pits—and an extensive cave system. Patient, thorough work in Jewel Cave, South Dakota, gave truth to the legends of a gigantic cave there. Explorers in Puerto Rico traversed an unknown underground river in rubber boats. The mysterious underwater caves of the Bahamas, the Blue Holes, were finally explored by scuba divers.

Other stories tell of the important role caves have played in American history and folklore since colonial days. There are several chapters about Mammoth Cave history. Mummies found in caves of that area were sold, stolen, and exhibited for money. Stephen Bishop, a black slave who was a guide to Mammoth, became one of its foremost explorers, and was consulted by scholars and scientists for his wealth of knowledge and expertise. A few hundred feet from Mammoth Cave lies Sand Cave, where Floyd Collins died slowly and painfully to the accompaniment of raucous publicity.

Lester Howe, discoverer of New York's Howe Cavern, claimed to have found a cave more beautiful than his namesake cave. He called it the Garden of Eden Cave. Since he never revealed the location, nobody is sure which of many caves in the vicinity it is—or whether Howe really found such a cave after all. In Indiana, Wyandotte Cave is known for its large size and the humorous folklore in its history.

Many caves are famous for their beauty and grandeur, including Carlsbad Cavern with its cathedral-size rooms. Others, such as the little-known Caverns of Sonora, Texas, which have exquisite crystal formations, deserve greater fame than they now enjoy. There are caves in lava which are extensive and beautiful, even though they lack the splendid decorations of limestone caves, as the history and exploration of the Saddle Butte system of Oregon reveals.

Cave animals, many of them blind but with highly developed sense organs, have been studied for their special ecological adaptations underground. Some show explicitly the processes of evolution. You will read about caves in Arizona that were visited thousands of years ago by now-extinct giant ground sloths, whose droppings in the caves helped provide knowledge of the climate and vegetation of that time. Although caves are generally healthy places to visit, some can transmit histoplasmosis, a mysterious disease that doctors formerly confused with pneumonia.

Like much of the environment above ground, caves and the animals that inhabit them are threatened by pollution, vandalism, and the pressures of human use, and these concerns are expressed in many chapters. Some cave bats are facing extinction, although measures are being taken to see if their survival can be assured. In Missouri, the Ozark Underground Laboratory is a unique demonstration cave that concentrates on public training and awareness of cave ecology and conservation.

Types of Caves

Limestone Caves

Visitors to limestone commercial caves are often told that the cave was formed by an underground river. They may wonder how a river could produce passages that go up and down, get larger and smaller, have dead-end passages, and sometimes form intersecting networks. No surface stream behaves in such a fashion—and neither do underground rivers.

Caves *are* formed by water, however. Chemically, pure limestone is calcium carbonate, which can be dissolved by water. The process is vastly speeded up if the water contains some acid, and, as most groundwater absorbs some carbon dioxide from the air, soil, or impurities in the limestone itself, the groundwater can begin the process of solution. Other rocks, such as marble (which is also calcium carbonate), dolomite (calcium-magnesium carbonate), and gypsum (calcium sulfate), are also water soluble. Cave making, or speleogenesis, generally begins below the water table in what is known as the zone of saturation or phreatic zone. A part of the rock is dissolved, and a void or cave results. The solution is concentrated along joints or cracks in the limestone, and other areas of weakness, such as bedding planes, that the water can reach and dissolve. Since joints often develop in sets at or near right angles to each other, caves developed in such limestones often show patterns that resemble city streets. Sometimes the incipient cave at this time resembles a giant sponge or swiss cheese from random solution.

As cave making continues below the water table, erosion takes place on the surface. Master streams, which control the level of the groundwater, will erode and lower their base levels. The groundwater will flow toward the master stream, and the cave, as it grows larger, serves to channel more and more groundwater toward the stream. Sinkholes may develop on the surface as the ground collapses or the bedrock dissolves over the ever-enlarging cave system. Smaller surface streams, their flow pirated, can be diverted underground to become a part of the increasingly efficient hidden drainage system.

As the master stream lowers the base level, the water table in the area will also be lowered. The cave is slowly drained of water as it fills with air. This is often a time of collapse, as blocks of rock

A caver gazes at a massive stalagmite in New Cave, Carlsbad Caverns National Park, New Mexico. *National Park Service.*

fall from their own weight without the support of the water, building the impressive piles of breakdown in many caves. Vertical shafts may develop below sinkholes and down vertical joints, which serve to channel water underground. These shafts are called pits, domes, or domepits, depending on the caver's point of view as he or she encounters them.

This slender sodastraw stalactite is surrounded and encrusted by needle-like crystals of aragonite. *Skyline Caverns.*

The cave continues to function as a drainage system as surface streams find their way into it. The lowest stream level in the cave often corresponds with the local water table. Down below, new caves may be developing. Fluctuating water tables tend to develop caves with several levels of passages, particularly if the periods between the fluctuations are lengthy and conditions stable. Wet and dry climatic changes during Pleistocene glacial and interglacial times can be correlated with some multilevel development. Present-day sea levels are considerably higher than they were during glacial periods, and many caves in Florida, the Bahamas, and elsewhere now underwater give abundant evidence that they were at one time filled with air.

The speleogenesis of a large cave system depends on a propitious combination of factors—geologic, hydrologic, climatological, and others, including such diverse factors as the temperature and pressure gradient of the groundwater. It is not just chance, for instance, that the largest passages in Kentucky's Flint Mammoth Cave System lie under flat-topped ridges. The ridges are flat-topped because of an impervious bed of sandstone which helps channel and contain the groundwater, greatly increasing its cave-forming function. The freshwater springs of Missouri, of northern Florida, and of the Blue Holes of the Bahamas all attest to active speleogenesis. In West Virginia the largest caves seem to occur in the headwater areas of major stream systems, probably because erosion farther downstream has taken away most of the big caves there. Carlsbad's huge rooms indicate a lengthy period of stable or slowly subsiding water-table conditions—certainly different from the present-day desert climate.

Once a cave is filled with air, deposition of speleothems or formations—stalactites, stalagmites, and other decorations—can start. Most result when drops of water carrying calcium carbonate in solution deposit a small amount of calcite on the ceiling (stalactites) or floor (stalagmites). Flowstone results from mineral-saturated water flowing over the walls. Speleothems may be precipitated in standing pools of saturated water, resulting in what are called lily pads or rimstone pools. Many speleothems are given descriptive names, such as bacon, soda-

Stalactites hang down over more massive stalagmites in a grotto of the Big Room; Carlsbad Cavern, New Mexico. *National Park Service.*

straw stalactites, or curtains, which have nothing to do with their origin. Cave coral is tough on cavers' knees, but its name comes from its roughness, not from the way it was formed.

Some speleothems, such as helictites, grow in response to a random crystal orientation, causing bizarre shapes which seem to defy gravity. Gypsum speleothems result from impurities already present in the limestone which are extruded into the open void of the cave. Even ice speleothems are known. Perhaps the best semi-technical classification of speleothems (there is no non-technical classification) is that of William R. Halliday in *American Caves and Caving*.

Limestone is so widely distributed geographically that the majority of states have at least one cave. Some areas are particularly rich in caves, however. In the eastern United States numerous caves are found in the limestones along the breadth of the Appalachian Mountains, in the Interior Lowland Plateau, and in the Ozark Plateau stretching to the West. This indistinct region covers parts of Pennsylvania, Maryland, Virginia, West Virginia, Tennessee, Alabama, Kentucky, Indiana, Illinois, Mis-

One of the finest examples of a rimstone pool is in the Lilypad Room of Onandaga Cave, Missouri. *Howard N. Sloane.*

souri, Arkansas, and some neighboring states; it surpasses any other area in the world in the number, size, and density of its caves.

Other major limestone cave areas include the Black Hills of South Dakota, the Edwards Plateau of Texas, much of northern Florida, the mountains of Montana and Wyoming, and the Guadalupe Mountains of New Mexico and Texas. Many smaller areas of the western United States are also speleologically important, but the limestones are scattered, even though there are individually important caves or groups of caves.

Some individual caves seem to defy the rules, and appear where they might not be expected. Caves in Canada and much of northeastern United States tend to be small, since most have developed since the last retreat of the continental ice sheet. But the longest cave in Canada, Castleguard Cave, runs underneath a glacier in Banff National Park, Alberta. The cave is entirely in limestone, and its entrance lies beyond the edge of the glacier. But the end of the cave, six and a half miles in, is blocked by the Columbia icefield, which is estimated to be 1,100 feet thick at

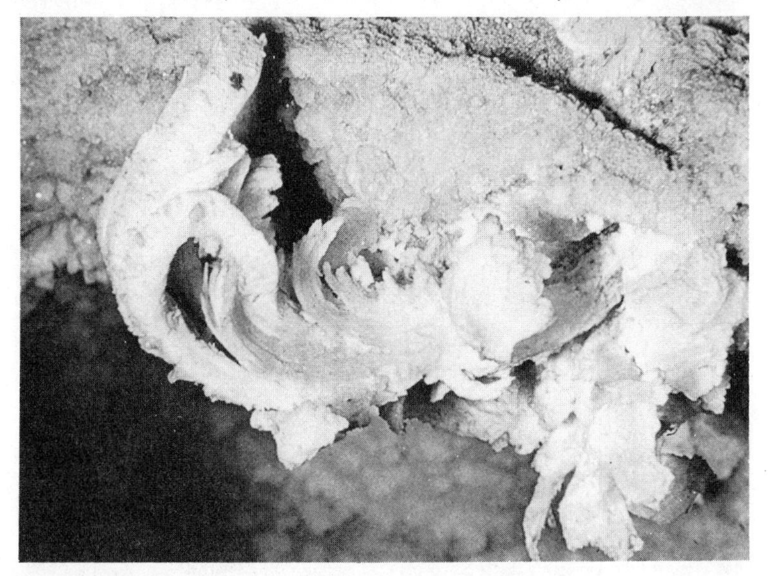

This gypsum crystal was extruded into the open void of the cave after the surrounding limestone was dissolved away. *Howard N. Sloane.*

this point, giving the speleologist a unique inside-out view of glacial activity.

Lava Tube Caves

Lava tube caves are born in fire as lava pours onto the surface of the earth. Only basaltic lavas—low in silica and high in iron and calcium—form caves, and then only under special circumstances. Lava is a viscous rock that flows downhill from its source, following low spots in the topography. The flowing lava cools quickly on its surface where crusts of semi-cooled lava form, although underneath it may still be molten. Downslope, at

An apparently random pattern of crystal growth has resulted in the bush-like appearance of this helictite in Wind Cave, South Dakota. The hand, lower right, gives scale. *Charlie and Jo Larson.*

High in the Montana Rockies a spelunker looks down on a valley from a cave entrance. *Bruce Sloane.*

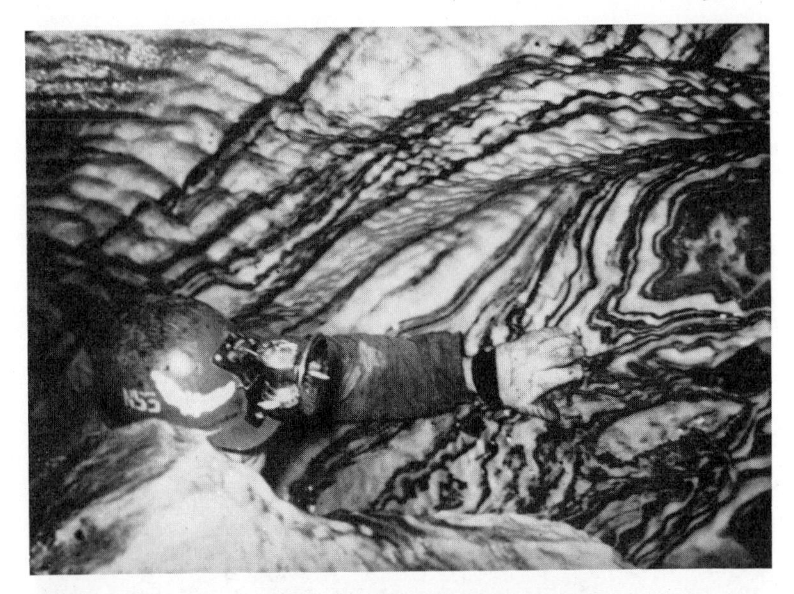

Caves in marble show features similar to those of limestone caves. This cave in Massachusetts shows beautiful banded walls. *Howard N. Sloane*.

Baker Cave, Oregon, is a lava tube cavern with large passages. A seasonal stream flows over the sandy floor. *Charlie and Jo Larson*.

the leading edge or toe, the hot inner core may break through the weak crust, and drain from the center of the flow. A hollow, a lava tube cave, is the result.

Most lava caves develop on gentle slopes of about one degree. If the slope is steep the lava flows too fast for a stable crust to form, since it is continually broken up and rafted downstream. Lavas on flat surfaces cool gradually and the inner core does not drain away. Caves are most apt to form in long, relatively narrow flows that permit rapid cooling of the overlying crust.

When the fluid basalt first reaches the earth's surface it is generally a ropy, smooth-flowing lava known by the Hawaiian term *pahoehoe* (pronounced "pa-ho-ee-ho-ee"). As the lava cools, it gives off quantities of gas, becoming less fluid and more blocky and clinker-like; it is then called *aa* (pronounced "ah-ah").

Most lava caves are simple in plan, being essentially straight

As this lava cave cooled, drops of molten rock splattered from above, forming lava stalagmites. Jo Larson admires these unusual speleothems in Lavacicle Cave, Oregon. *Charlie Larson.*

or gently meandering tubes. Occasionally multi-level caves form, or an earlier cave may be blocked by a later flow. Since the caves form on slopes, they have an up-tube and down-tube direction. Cave entrances, or sinks, are usually areas where the roof has collapsed. Since these caves develop close to the surface, some systems have numerous entrances. Often a large portion of the cave collapses as it forms, resulting in a lava trench which may be quite extensive.

Lava flows are fairly common geological events. Although lava may accompany explosive volcanic eruptions, more often it flows out quietly from cinder cones or volcanic vents. Lava caves can be found wherever there has been recent volcanic activity, including many islands throughout the world, and in many parts of the western United States. They are among the most recently formed geologic features; observers in Hawaii have seen lava

Rough and jagged *aa* lava fills and blocks the passage in Spider Cave, Washington. The walls and ceiling are enclosed by smoother *pahoehoe* lava. *Charlie and Jo Larson.*

tubes forming through gaps in crusted-over flows.

Unlike limestone caves, lava tubes do not change much once they have formed. Some speleothems can result as the molten rock cools and drips, and later weathering may produce bizarre shapes. Lava tube caves are relatively short lived, either being filled or covered by later flows, or collapsing within a few thousand years.

Glacier Caves

Caves in glacial ice have been known and visited on the flanks of Mount Rainier, Washington, for many years, but until recently received little attention. They were thought to be seasonal features enlarged by summer streams and then destroyed by glacial flow with the advent of winter. Explorations now indicate that some of the caves are quite extensive, with some beautiful ice speleothems. Most of the glacier caves on Mount Rainier are formed by streams, and their passages increase in size during the summer months from ablation (evaporation) of the glacial ice.

Heavy snows during several winters of the early 1970s reversed at least temporarily the general retreat and shrinking of Mount Rainier's glaciers. The snow did not melt entirely over the summers, but became compacted and more granular. Such snow is known as "firn," and, with continued compaction, it becomes a part of, and indistinguishable from, glacial ice. New caves contiguous with the old developed in the firn ice, and by the mid-1970s more than ten miles of cave passages had been mapped beneath Mount Rainier's glaciers. Whether the increased snowfall represents a significant climatic change, in which case the caves can be expected to grow and enlarge along with the glaciers, or whether it is a short-term minor fluctuation, remains to be seen.

Renewed interest in Mount Rainier's glacier caves has stimulated exploration for glacier caves in general. The most promising areas in North America lie beneath the glaciers of Alaska and western Canada, and several caves have been found. They require special techniques—a combination of winter mountaineering and speleology. Some glacier caves are better explored in winter when the dangers of ice fall or a sudden rise in stream level are at a minimum, but the threat of hypothermia (loss of body heat through exposure) is real, and serious. Glaciospeleology is in its infancy, and there is still much to be discovered.

Sea Caves

Sea caves, or more properly, littoral caves (littoral refers to the beach or coast) are carved by the erosive action of waves at the edge of the sea or of a large lake. These caves can form in virtually any kind of rock along the edge of any body of water large enough to produce waves. The waves at first enlarge zones of weakness in the rock, which may be a crack or joint, or an area of weaker rock. Once an incipient channel is worn away, the battering action of the water may act like a powerful piston

A stream flows out of the Paradise Ice Caves beneath a glacier in Mount Rainier National Park, Washington. The ceiling flutes are caused by evaporation of the ice by warm air. *National Park Service.*

eroding and corroding the cave, which may reach a length of several hundred feet, although most are much shorter.

Sea caves are common along both the Atlantic and Pacific coasts of the United States and Canada wherever the coastline is made up of bedrock. Anemone Cave in Acadia National Park, Maine, is perhaps the best known. Oregon's Sea Lion Cave is home to hundreds of seals, one of the more unusual cave-utilizing animals. In the Rocky Mountain states small sea caves formed at the edge of many freshwater Pleistocene lakes which

An entrance to the Paradise Ice Caves as it appeared in 1950. Heavy snows in the early 1970s have made the caves almost inaccessible. *National Park Service.*

are now dry. Of course sea caves can form in many different kinds of rock, and these differ considerably from now-flooded solution caves—almost always formed in limestone—in coastal areas.

Because sea caves are continually exposed to hydraulic battering-ram wave action, they often contain loose rocks ready to be dislodged by the next wave or caver. Some of them may be underwater at high tide, and others have floors covered by ubiquitous slippery marine growth.

Talus and Boulder Caves

Talus and boulder caves form in mountainous areas. Water works its way into cracks in the rocks. When the water freezes into ice it expands, exerting great pressure. Alternating melting and refreezing eventually dislodges large and small blocks which can accumulate in huge piles at the base of cliffs, forming what is called talus. In glaciated areas, which includes almost all the mountains of North America, glacial action may have helped open up cracks and joints in the rock, and the glaciers moved

Anemone Cave, Acadia National Park, Maine, a sea cave overlooking the Atlantic Ocean, can be entered only during low tide. *Charles E. Mohr.*

rocks around or piled them up.

Some individual boulders in talus piles can be quite large, tens of feet long. There may be spaces large enough for people to enter. Although most of these caves are small and never get completely away from daylight, some extend for surprisingly long distances, hundreds or even thousands of feet, often in complete darkness. New Hampshire has a talus cave with over three thousand feet of passage—no Carlsbad—but many more exist which haven't been adequately investigated. Some in New Hampshire have been developed as tourist attractions. Washington, and other western mountain states, also have many talus caves.

Miscellaneous Types of Caves

Rock shelters are cliffs or steep slopes whose upper parts overhang the bottom portion.. They are not usually considered true caves since they do not extend into darkness, although many are quite scenic and have important archaeological or biological significance. Correctly or incorrectly, most are erroneously referred to as caves, including many containing spectacu-

A young woman leaves one of the talus or boulder caves at New Hampshire's Polar Caves. *Dick Hamilton, courtesy White Mountains News Bureau.*

Above: A caver checks his camera before entering this crevice-like fault cave in Pennsylvania. The cave is the result of earth movement, which separated the two sides of the crevice. *George F. Jackson.*

Right: Crystal Ice Cave, Idaho, a lava fissure cave. Year-round ice is found only in caves with poor air circulation in cold climates. Cold, dense air, which enters during winter, remains year round and ice speleothems may result. Solution caves in limestone usually have good air circulation and rarely have ice speleothems. *James L. Papadakis.*

lar and interesting cliff dwellings in the southwestern United States.

A fault is a fracture zone in the earth's surface whose opposite sides have moved relative to each other. Often there are spaces large enough to enter between the two sides of the fault. Limestone solution caves sometimes follow fault zones since these may be zones of weakness allowing groundwater flow.

Volcanic rift zones are fissures on the sides of volcanoes. Crystal Ice Cave in Idaho is developed in such a fissure; as its name suggests, spectacular ice speleothems decorate this cave.

A variety of minor processes can result in caves, some of which may be locally important. Blocks of rock detach from cliffs, producing fissure or block creep caves. In time, as the block creeps downslope, fissure caves grade into talus caves. In Yellowstone National Park and other thermal areas, travertine deposits form small caves, and the abandoned now-dry plumbing of former hot springs and other thermal features can occasionally be entered. Streams sometimes erode channels and undercut friable bedding planes for short distances. But most of these have little significance.

The Flint Mammoth Cave System

The Mammoth Cave area of Kentucky is unique. Under its ridges and valleys lies the world's longest cave—the Flint Mammoth Cave System—with about 180 miles of surveyed passage. As mentioned, its human history is as complex as the cave maps. It goes back to pre-Columbian man, and encompasses the life works of scores of individuals, including such speleologists as Stephen Bishop, Floyd Collins, and James W. Dyer. The animal life of the cave is as varied and rich as that known anywhere else. So much has been written about this cave system, and so much still needs to be said, that it has been difficult to restrict selections for this book.

The Flint Mammoth Cave System is made up of five caves originally thought to be separate and distinct: Mammoth Cave, Floyd Collins' Crystal Cave, Colossal Cave, Salts Cave, and Unknown Cave. Mammoth Cave lies under a flat-topped ridge called Mammoth Cave Ridge. To the northeast is Flint Ridge under which the other four caves lie. The ridges are separated by the steep-sided Houchins Valley.

Written accounts of exploration of the caves date from the 1790s. Mammoth Cave was important as a source of saltpeter, or potassium nitrate, the main ingredient in the manufacture of explosives and gunpowder during the War of 1812. But the greatest development occurred following the Civil War, as the cave tourist business attracted commercial interests, through the 1930s. Several caves without natural entrances were discovered, including Crystal Cave, discovered in 1917 by Floyd Collins, who dug in the bottom of a sinkhole. The names of many of the early explorers, including Bishop, Hunt, Collins and others, can be seen in the caves today, but most of them left no written records of their work.

Extensive ice speleothems have completely blocked the passage in this lava tube. *Charlie and Jo Larson.*

Guides continued to explore and extend the known boundaries of the caves during the 1930s and 1940s. Mammoth Cave was established as a national park in 1936. In 1938 the "New Discovery" section of Mammoth Cave was found by guides Pete Hanson and Leo Hunt. In 1947 James W. Dyer, manager of Floyd Collins' Crystal Cave, began systematic exploration there. A week-long expedition by sixty-four people in 1954, sponsored by the National Speleological Society, consolidated what was then known about Crystal Cave. This led to the formation of the Cave Research Foundation in 1957, and most work in the area since then has been under its direction. The four large Flint Ridge caves—Crystal, Colossal, Salts, and Unknown were soon linked together.

But Mammoth Cave still lay more than a mile away across Houchins Valley. Passages extending south from Flint Ridge into the valley were narrow, sinuous crawlways, many almost

Explorers forced their way down Hanson's Lost River to connect the Flint Ridge Cave System to Mammoth Cave on September 9, 1972. *Copyright © 1972, Cave Research Foundation. Roger W. Brucker.*

filled with mud and water, and subject to periodic flooding. By 1965 exploration finally passed completely beneath Houchins Valley, 300 feet under Mammoth Cave Ridge. But explorers were stopped by massive breakdown, and could not find a way through.

The breakthrough finally came in 1972 under the leadership of John P. Wilcox, who staged a major assault. A survey team headed by Patricia Crowther negotiated many tight passages under Houchins Valley, and discovered a stream passage heading toward Mammoth Cave. On a wall, team members Tom Brucker and Richard Zopf saw the scratched letters "PETE H" with an arrow pointing *away* from Flint Ridge. "PETE H" obviously was Pete Hanson, who must have come from Mammoth Cave, and his arrows always pointed *out*. The connection must be close. But the party, tired by a sixteen-hour trip, and facing

John P. Wilcox knew his party had connected the Flint Ridge Cave System with Mammoth Cave when he spotted a steel handrail. *Copyright © 1972, Cave Research Foundation. Roger W. Brucker.*

the prospect of a return trip of several more hours, had to turn back.

On September 9, 1972, Wilcox was back with a team of the slimmest and strongest cavers he could summon. In addition to Wilcox, Crowther, and Zopf, the party consisted of P. Gary Eller, Steve G. Wells, and park ranger Cleveland F. Pinnix. They entered the cave at 10:00 A.M. and quickly threaded their way back to Hanson's signature. Pushing on, they mapped nearly a mile of additional passage. The water here reached nearly to the ceiling, and further exploration would mean complete immersion and chilling, seven miles away from where they entered. Wilcox did not hesitate. A few minutes before midnight he slipped into the water and passed through the deep portion. The cave opened up, and his lights revealed a steel railing and walkway beyond. This must be Mammoth Cave! "I see a tourist trail!" he shouted, and the excited party pushed after him through the water.

The old tourist trail in Mammoth Cave's Cascade Hall passes by the low opening (lighted) of Hanson's Lost River. *Copyright © 1972, Cave Research Foundation. Roger W. Brucker.*

They were indeed in Mammoth Cave, in Cascade Hall at the end of the old commercial boat trip, no longer visited by the public. But thousands of tourists had passed by this very spot. After the excitement subsided the group re-entered the water to unite the survey of Mammoth Cave with the Flint Ridge Cave System. Now the world's longest cave was the Flint Mammoth Cave System.

The breakthrough was not entirely unexpected. Wilcox had a sketch map in his pocket showing a possible connection at about this point. And the long trip back was not necessary; Ranger Pinnix produced a key to Mammoth Cave, and the party, still exhilarated, left via the elevator at the Snowball Dining Room, completing the first portal-to-portal traverse from Flint Ridge to Mammoth. What seemed surprising was that nobody in recent times had explored that passage, now known as Hanson's Lost River. Several old maps indicated a short dead-end passage from Cascade Hall. But most of the time the passage was underwater. Only during very dry summers—such as the summer of 1972—was the water low enough to permit entry.

The discovery united the Flint Ridge Cave System, 86.5 miles in length, with Mammoth Cave's 57.9 miles of passage, producing a combined length of 144.4 miles (232.5 kilometers). Additional explorations are continuing at the rate of about ten miles a year. Other caves in Flint Ridge may eventually be found to connect with this system. Crawlways from Mammoth Cave have reached within a few hundred feet of Sand Cave, where Floyd Collins met his death. But the greatest possibilities lie to the south under Joppa Ridge, which, like Flint Ridge, is separated from Mammoth Cave Ridge by a deep valley. Several caves are already known in Joppa Ridge, including Lee Cave (7.6 miles long) and Proctor Cave (5.5 miles). Someday speleologists will find their way from these caves to connect with the Flint Mammoth Cave System. The passages are there: they need only to be discovered by humans.

A complete history of Mammoth Cave and Flint Ridge would fill a library. The story of the 1954 Crystal Cave expedition is told by Joe Lawrence, Jr., and Roger W. Brucker in *The Caves Beyond*. The most accurate and complete account of Flint Mammoth exploration is *The Longest Cave*, by Roger W. Brucker and Richard A. Watson of the Cave Research Foundation.

Cave Life

Just about any animal can be found in cave entrances. Spelunking cows have been encountered 400 feet inside some West Virginia caves on hot summer days, and Texas cavers have been known to wait hours inside a cave for rattlesnakes to vacate the premises.

But these are accidental visitors. Caves provide specialized environments. Not only is it totally dark, but food is sparse. The usual diurnal rhythm of night and day is missing, and there are few, if any, seasonal changes. Animals that adapt to such condi-

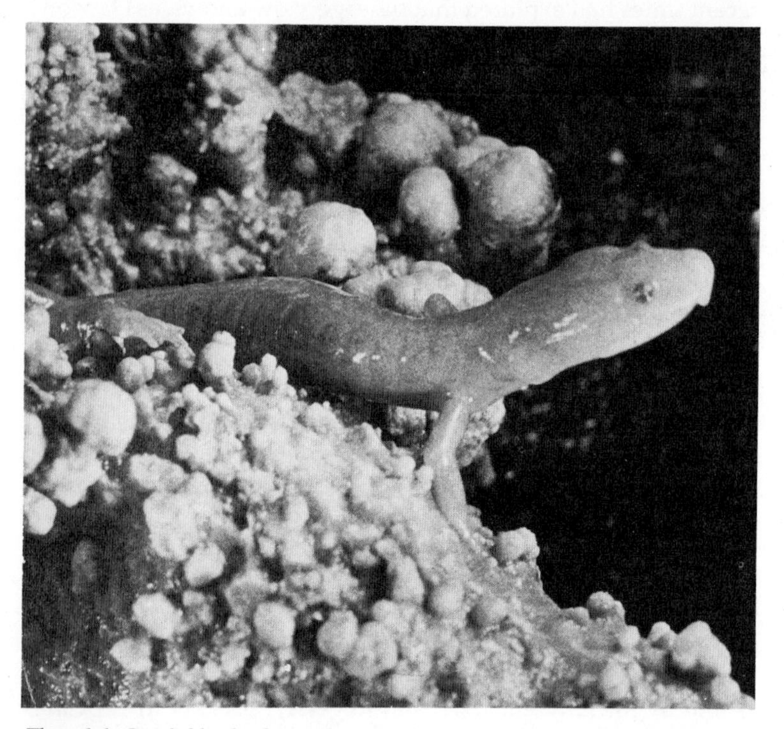

The adult Ozark blind salamander, *Typhlotriton spelaeus*, has degenerate eyes. Totally blind, it stalks its prey by touch and vibration. *Charles E. Mohr.*

tions must themselves become highly specialized, a process that takes generations of evolutionary adaptation. If something should then happen to change the environment, the animal may be unable to adjust, and may face extinction or extreme reductions in population. Unfortunately, since the activities of humans do change the environment, and rapidly, this situation is already a reality for many cave-dwelling creatures.

Several chapters in the book deal with cave organisms and the problems that confront them. Biologist Thomas C. Barr, Jr., discusses the evolution and adaptation of salamanders, insects, fish, and crayfish in the Mammoth Cave area. In "Refugees of the

This blind crayfish from Wyandotte Cave, Indiana, uses touch and feel to find food. *George F. Jackson.*

Ice Age" he tells how these creatures, many of them eyeless, survive and adapt in their subterranean habitats. Bats are also victims of ecological change. Although bats find their food outside caves, several species seek out caves close to their food source, or hibernate in caves with their precise requirements of temperature and humidity. Any disturbance during hibernation may deplete their energy reserves drastically. Some species of bats have already been greatly reduced in numbers. Naturalist Charles E. Mohr discusses two endangered species in "Survival: A Tale of Two Bats."

The food supply and food chain are vital factors in the survival of an animal. Tom Aley tells how his research cave, the Ozark Underground Laboratory, has helped speleologists to understand the food chain and the interrelationships between the surface and the cave below. His laboratory cave also serves as a refuge for several endangered species.

Visiting Caves

Exploring undeveloped caves is not usually dangerous or hazardous for experienced cavers with proper equipment. Caving requires physical fitness, specialized equipment, skills, and experience. Most accidents happen to unprepared neophytes. In 1975 a couple almost starved to death in a Maryland cave when their only flashlight malfunctioned. They told nobody where they were going, not even the cave owner, and were not missed for several days. They were finally rescued after some ten days in the dark, with no food and only cave water to drink. Both had lost more than twenty pounds, and required hospitalization. If their pick-up truck had not been noticed parked near the cave, they doubtless would have died of starvation. Cavers frequently are called upon to rescue people "lost" in nearby caves; what they usually find are several sheepish, scared individuals waiting in darkness after the failure of their single flashlight.

There are other reasons not to visit caves. Any cave belongs to somebody, and the owner may not want you to visit his cave. A property owner has the legal right to refuse you permission—just as you may refuse a stranger the right to walk through your house. Often a cave owner will grant experienced cavers permission to visit, but, fearing accidents, lawsuits, vandalism to the cave, or other calamities, he may not permit others. Many caves owned by federal or state governments are also restricted. Some cave owners have been so disturbed by visitors who ring doorbells at 3:00 A.M., leave cattle gates open, carry off stalactites, and indulge in other undesirable behavior that they have installed locked gates at cave entrances, or even bulldozed caves shut.

For the sake of preservation, undeveloped caves should not be visited. Caves are fragile, containing some of the most delicate natural features on earth. A forest destroyed by fire can

A National Park Service ranger shows two visitors the Frozen Waterfall and other speleothems in the Green Lake Room of Carlsbad Cavern. *National Park Service.*

Ice speleothems, including several columns, remain all year in several California lava tube caverns. Cavers use extra caution around them, as they are more fragile than rock speleothems. *Charlie and Jo Larson.*

eventually be recovered. But a broken helictite is gone forever. Few caves have speleothems that are still forming. The rate of growth of those that are active is slow—so slow it is doubtful if your children's children will ever see more than a stub where once-grand stalagmites have been broken off. The federal government and many states now have strict laws protecting caves and their features. Even well-meaning cavers and speleologists who would never vandalize a cave may inadvertently destroy a delicate feature. Every experienced caver knows of too many once-lovely caves whose beauty has been destroyed by man.

The life in caves is fragile, too. Several once-abundant species of bats are now threatened with extinction. Other forms of cave life are found in small numbers in only one or a few caves, and could easily be wiped out completely. Speleologists try to avoid disturbing cave life, but often the mere intrusion of people can upset the delicate balance. They now try to avoid hibernating bats by closing some caves for several months each year to protect them.

Caves can be visited safely, and with a minimal amount of environmental harm, however; after reading some of this book, you might want to visit one. Many caves mentioned in the book are commercially developed, and are equipped with lights, walkways, and guides or interpretive services. Commercial caves include many of the largest, most interesting, and beautiful caves in the United States. Commercializing a cave may also help to protect it from vandalism. Few areas of the country are more than a day's travel from a commercial cave.

If you want to visit a cave, and I hope you will, first visit a commercial or show cave. Perhaps you should visit several. If you still want to go caving after this, write to the National Speleological Society. The address is given below. The Society will give you more information, and will help you to get in touch with local cavers.

The National Speleological Society

The National Speleological Society is the main nationwide caving organization in the United States, and is recognized world wide. Nearly all contributors to this book as well as its editor are NSS members, and many have made significant contributions to speleology. The Society is made up of both cavers and speleologists, but many place themselves in both categories. There are numerous local chapters, called Grottos, throughout the country. The Society publishes several periodicals, including the monthly *NSS News*, and sponsors an annual convention. In addition to scientific endeavors, the NSS concerns itself with underground safety, conservation, land-owner relations, and other speleological matters. The address of the Society is:

The National Speleological Society
Cave Avenue
Huntsville, Alabama 35810

Chasing the Winds Through Jewel Cave

The wind blew through the cave passage. It must come from someplace—so thought Herb and Jan Conn, husband-and-wife exploring team, as they made their way along the corridors of Jewel Cave beneath the Black Hills of South Dakota.

Jewel Cave, a National Monument, had been open to the public for many years. But it had always been overshadowed by its neighbor, Wind Cave National Park. There were even rumors that Jewel would be removed from National Monument status.

But the Conns persisted. They followed the underground breeze through crevice and crack. After several trips the Conns found themselves in virgin cave—where nobody had been before—and they knew that Jewel was much more extensive than anyone realized. For more than ten years they explored patiently, enlarging the maze-like map until it recorded more than fifty miles of underground wonders. They tell their complete story in their book, *The Jewel Cave Adventure.*

Today Jewel is recognized as one of the largest caves in the United States, far longer than its neighbor Wind Cave, and explorations are still continuing. A new entrance and elevator make some of the remote areas explored by the Conns accessible to the public. And there is no more talk about removing Jewel from the National Park system.

Chasing the Winds Through Jewel Cave

HERB AND JAN CONN

In 1959 a sign at the entrance to Jewel Cave said, "This is a small cave, but the tour is as exciting as an exploring trip. You will traverse steep stairways and low narrow fissures."

Fine, we thought. A small but exciting cave should be ideal to introduce us to cave exploring and mapping. We didn't want to tackle too large a cave as a first assignment. When we finished with Jewel, if we were still enthusiastic, perhaps we could find a bigger one.

Dwight Deal had invited us to join him in this project of cave surveying and mapping. Jewel Cave is a National Monument fourteen miles west of Custer, South Dakota, in the Black Hills where we had lived for eight years, largely ignorant of what lay beneath the surface. Dwight had been granted permission by the National Park Service, which administers Jewel Cave and conducts guided tours through its passages, to explore beyond the developed trails for the purpose of preparing a new and more extensive map of the cave. So we were hopeful that we would get

---

Beyond the Long Winded Passage the cavers climb over breakdown, blocks of fallen rock. *Dave Schnute.*

to see more than the steep stairways and narrow fissures of the conducted tour.

Some people had been speculating ever since its discovery that Jewel Cave might be more than a "small cave." The Michaud brothers, Albert and Frank, came upon the entrance in 1900, a small black hole on the side of Hell Canyon with an awesome wind issuing from its depths. They were convinced they had found the other end of Wind Cave, over twenty miles away. Any cave spanning twenty linear miles of countryside is not a small cave. But it seemed logical that the strong wind known to blow in and out of Wind Cave must find a route to the surface somewhere else. When millions of cubic feet of air blow into a hole in the ground, it's reasonable to suppose that some other hole somewhere is blowing out.

After enlarging the entrance, originally too small for a person to enter, the Michauds followed the Jewel Cave wind through tortuous crawlways and walkways into blackness where no light had been before. In the glow of their candles they could see calcite crystals, the jewels for which the cave was named. They found walls covered with a crystal layer two to eight inches thick, slabs of fallen crystal underfoot, crystal everywhere they looked. Most of it was dull and stained with dirt, but occasional pockets of clear sharp-pointed crystals were jewel-like indeed.

Tackling the job of cave exploration with enthusiasm, Albert and Frank Michaud brought in ropes to descend into the pits where the cave wind lured them. The left fork of the cave led at last to the Dungeon, a room 700 feet from the entrance and ninety feet below it, where water dripped sporadically and bats roosted on the ceiling. Hunting for a continuation of the wind, they climbed a wide slippery chimney back to the cave's upper levels, but in the maze of winding crawlways and breakdown-choked passages beyond, the wind eluded them.

In the right fork of the cave the wind was stronger than in the left, but exploration there was postponed because the drop-offs were more frightening. After building a ladder to descend the first sixty-foot pit, they followed the lower passage to a second drop of fifty feet. Another ladder was required to plumb this one. From the Heavenly Room beneath the second drop they needed a third ladder, this one to climb up into the continuing passage. Wind raced along at a merry clip to Milk River, a wet little grotto where water dripped from a host of small stalactites onto a mud slope, staining it white before oozing on into a pool.

The Milk River stalactites have been broken now, by careless visitors who struck them by accident, or by thoughtless visitors who wanted a souvenir. Thousands of visitors have been to the Dungeon and to Milk River, guided in the early years by the Michauds, later by National Park Service rangers. Not so many, Dwight Deal told us on our first trip, had been beyond Milk River.

It was easy to see why. To continue we had to get down on our stomachs in the wet mud and slither, for the ceiling height was only about one foot. Once we were well dunked in mud we found the passage had become dry again, but it remained uncomfortably cramped for the next 500 feet.

Herb and Jan Conn at Milk River in Jewel Cave on one of their many exploring trips. *Dave Schnute.*

Somewhere in the Badger Hole Traverse, rangers Ryan, Hylton, and Kinzer had made a startling breakthrough the previous year. Noticing air movement into breakdown, they had dug the passage open and continued to an impressive brink, looking off into a chasm fifty feet deep where their lights were swallowed in the blackness. After 500 feet of crawling this new passage looked especially inviting—but how did one reach it? The floor was thirty feet straight down.

With a caver's ladder of aluminum rungs strung between steel wires, the rangers subsequently descended to the floor of the "Discovery Room" and explored a quarter mile of spacious, virgin passage. Delmer Brown, working on his bachelor's thesis at the South Dakota School of Mines, wrestled a surveyor's transit through the tight crawlway, surveying from Milk River to the thirty-foot drop. Bill Eibert, using the less accurate but more convenient compass and steel measuring tape, continued the survey through the larger passages beyond.

Eibert's survey ended in a nook Dwight called Duffy's, lunchroom of the underground élite. Explorers had been a little farther, but not much. After we had made an airy crossing of Bell Bottom Canyon, between flaring crystal walls, there were no footprints ahead of us.

Our first day in the cave was a good one. A good day is one when we peer into more open passageways than we have time to follow. We go home feeling bewildered by the complexity of what must lie ahead, and we're eager to return to solve the mysteries. A bad day, on the other hand, is when we start poking all those good leads, and each of them stops except two—and these prove to be connected, so that we enter one and emerge a few minutes later from the other.

The cave was still opening ahead of us, and we were still eager to push on when Dwight announced it was time to go back and survey what we'd already found. Turning back was hard, but there was much to do. As we retreated, however, we learned

---

Herb Conn carefully makes his way across crystal-lined Bell Bottom Canyon. *South Dakota Highway Department.*

that it isn't prudent to extend oneself too far. Already we had forgotten some of the turns we had made.

Fortunately Dwight remembered the way better than we did. He taught us the rules of sensible caving along with the technique of surveying with hand-held compass, clinometer, and tape. Primarily, he explained, caving must be *fun*. But it isn't fun if you're lost. It isn't fun if you're hurt. And it isn't fun if you drop your carbide lamp down a hole and then discover that your flashlight batteries are dead.

We found, too, that it's more fun when you map conscientiously as you go. Explored passages that aren't mapped are soon forgotten and have to be explored again. Unless we survey, we have nothing to show for our day when we get home. Adding the new passages to the map—seeing how far they go, how they fit together—is a satisfaction that makes surveying well worth the effort.

After we had extended what seemed to be the main line of the cave 1,200 feet beyond Duffy's, east under the hill bordering Hell Canyon, we came to a dead end. There were still lots of side leads to check. One side lead took us up past a beautiful display of aragonite "frostwork" into a loft, the highest cave level where

Jan Conn reads the survey tape at Station L4. *Herb Conn.*

walls are apt to be bare limestone, free of crystal, but fantastically sculptured into alcoves, arches, and domes.

There was a side lead to the north way back in the Badger Hole Traverse, just before the location of the 1958 breakthrough. Lots of people had already tried it; in fact, the Michauds had often conducted cave visitors as far as a large boomerang-shaped room in this area. In those days, we suppose, people were hardier. The Milk River mud was no worse than the mud travelers found along the road, and slithering on their stomachs through the crawlways was, perhaps, a welcome change from bouncing in a buckboard wagon.

Dave Schnute joined our survey team during the second month. Dave is a natural lover of caves, a friend to the crystals and bats and the damp, chill darkness. He found the way down through a jumble of fallen blocks east of the Boomerang Room to a passage at a lower level. The floor seemed to be untracked. Confident we had located another new branch of the cave, we hurried on east. Leading, Dave ground to a halt in a tight place.

We encouraged him to fight his way through, secretly hoping, however, it would dead-end so we wouldn't have to follow. No such luck. Dave emerged into the Gear Box, the biggest room we had seen in the cave.

The Gear Box is shaped roughly like the letter *H*, reminding us of the manual shift pattern of an automobile transmission. Coming in we had been in low gear. Straight ahead the reverse gear arm stretched away into blackness, sixty feet high, impressive and vast. To the right, through the Neutral Bar, the other half of the room was equally spectacular.

More amazing, however, than these new leads to check was the beaten trail of footprints across the mud floor. People had been here, where we thought we were the first. Who? When? Had they been as impressed with this room as we were, and if so why hadn't they told others about it?

The mystery deepened when the tracks led us to the rotted remains of a rope. This rope, it appeared, had once served as a handline for a short climb. Nearby were other artifacts: a candle stub, a metal spike, a ball of mud molded by human hands, and a sheaf of pages from a Sears, Roebuck farm catalogue advertising

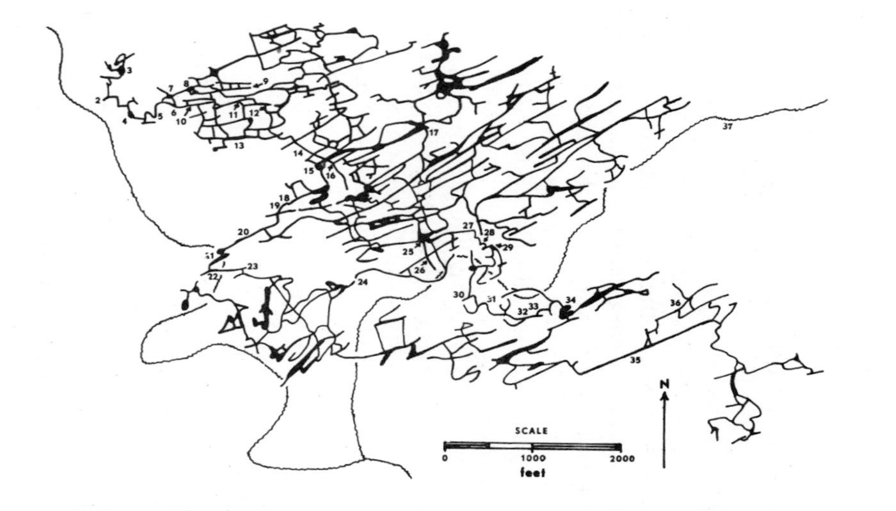

A much simplified map of Jewel Cave. 1. Hell Canyon; 2. Natural Entrance; 3. Dungeon; 4. Heavenly Room; 5. Milk River; 6. Badger Hole Traverse; 7. Boomerang Room; 8. Gear Box; 9. Forgotten Passage; 10. Discovery Room; 11. Duffy's; 12. Bell Bottom Canyon; 13. Eerie Boulevard; 14. Long Winded Passage; 15. King Kong's Cage; 16. Hollow stalagmites; 17. The Hub; 18. Tenderizers; 19. Snow Blower; 20. Small Favor; 21. Hell's Half Acre; 22. The Horn; 23. High Water; 24. Plug Ugly; 25. Target Room and Elevator; 26. Tunnel; 27. Hydromagnesite balloons; 28. Hurricane Corner; 29. Cloudy Sky Room; 30. Delicate Arch Room; 31. Humdinger; 32. The Miseries; 33. Calorie Counter; 34. Metrecal Cavern; 35. Mind Blower; 36. Gypsum flowers and beards; 37. Lithograph Canyon. *Map courtesy Herb Conn.*

horse-drawn plows. Whoever had left these relics, they were evidently old-timers. Had they for some reason kept their discovery a secret? Or—eerie thought—might we find their bones at the end of the trail of footprints?

In correspondence with Sears, Roebuck and Company the National Park Service determined the catalogue date as 1908. A tobacco can, subsequently found in another passage from the Boomerang Room, identified the route used by the early explorers. Some smoked initials suggested members of the Michaud family. And there *had* been rumors, it seemed, of a vast room lost somewhere in the back regions of the cave. A few returning visitors were insistent that they had been taken there by the cave guides of an earlier era. Lyle Linch, custodian-guide at Jewel in 1946-1947, tells us he was probably there, but at the time no one believed his "big room" story. Cavers and fishermen, perhaps, share a common urge to exaggerate.

After mapping everything we could find in this section north of the Discovery Room, in May 1961 we tried a small hole leading south. Here the soft dirt was untracked. No one had been before us. It opened up miles of cave which took us years to explore, a wilderness which never once betrayed a sign of human visitation. At times, as we pushed ever farther from home base, we longed to find a footprint or a gum wrapper, some friendly sign saying there might be, just ahead, a shortcut home.

After some pleasant meanderings through small upper-level passages we looked off into the mysterious blackness of Eerie Boulevard. When we had surveyed this new find, we had to tack on a southern extension to our map, a business that was to happen again and again in the following years.

The National Park Service personnel at the cave were excited by the recent discoveries. We remember with a special thrill the day Chief Ranger Dick Hart called the staff together and slowly began to unroll the new cave map. "This is all that was known two years ago," he explained, "but now" (there were gasps from the audience as he unrolled the rest) "look what we have!"

Proud as he was of the new extent of the cave, Dick was concerned that nothing we had yet found could be shown to cave visitors. With the advent of fine paved roads, smoothly riding

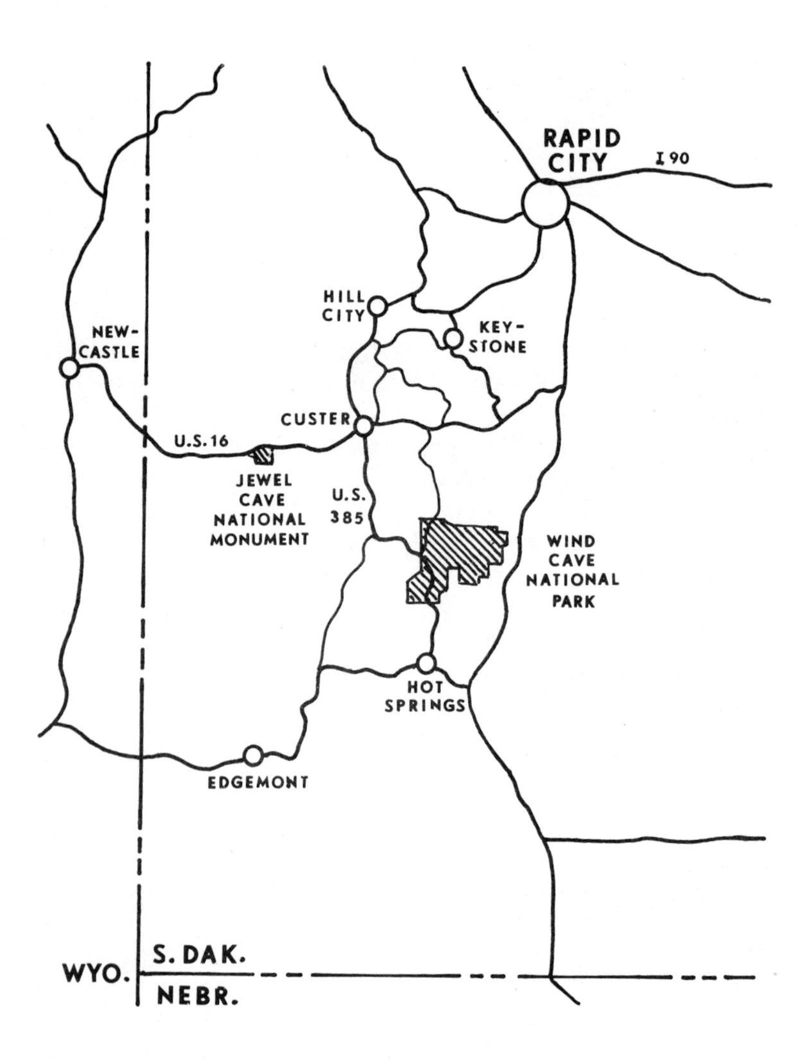

Jewel Cave is located near Custer in the Black Hills of South Dakota. *Map courtesy Herb Conn.*

automobiles, and comfortable motels, travelers did not come prepared for the long muddy crawl through the Badger Hole. The present tours, up and down the steep wooden stairs to Dungeon and Heavenly Room, were quite strenuous enough for the average visitor, more than enough for many.

Too, the narrow cave trail was at times overcrowded, especially as entering tours usually had to pass other groups on their way out. The cave itself, some Park Service officials thought, was not spectacular enough to warrant showing to the public as a government-sponsored attraction. Others, like Dick, thought that the tour was worthwhile as an adventure in caving under fairly primitive conditions (hand-held lanterns for light, watch your footing carefully, and don't bump your head!). But in our explorations, Dick suggested, we should give first priority to finding something else to show visitors, hopefully larger passages scenically decorated and on the *near* side of the Badger Hole. Such a discovery would both ease the overcrowding and still the critics.

For our own benefit and comfort we would have been de-

Sparkling calcite crystals—some "jewels" from Jewel Cave—grow from the wall. *Dave Schnute.*

lighted to find some other way to go than through the Badger Hole. But it was not to be. All other leads from the tour area were miserably tight and didn't go far. All the good stuff was out beyond the long crawl.

To make matters worse, soon we had another and tighter crawl to negotiate each trip. Air movement in Eerie Boulevard led us eventually to the Long Winded Passage, a narrow corridor trending southeast which is a comfortable two feet high at its start but gradually decreases to a minimum of 9½ inches.

The cave wind had dissipated in the large branching passages beyond the Badger Hole, but here it was again, not as strong as near the entrance but enough to flicker the flames of our carbide lamps. Blowing hardest at the tight spots, it waits to blow out our lights until we're caught in a squeezeway with arms above our heads, helpless to hunt for an alternate source of light.

Wind is a sign that the passage goes somewhere and that there may not be many other ways to get there. This proved true of the Long Winded Passage. In the twelve years since our breakthrough into the larger passages beyond in March 1962, we've found but one alternate route, and it is more devious, every bit as tight, and requires a rope and climbing skill to negotiate.

When the Long Winded Passage finally opens out, it does so in a grand manner. Pete Robinson, new head ranger at the cave, was the first to gaze off into King Kong's Cage, a room surpassing in volume any known previously in Jewel Cave. From here the passage is spacious all the way to the Hub, a quarter of a mile to the east, and similar parallel passages span the same interval one after another on to the south.

Our mapping efforts could not keep pace with the rate at which the cave unfolded. For years we passed great yawning halls we had never taken the time to enter.

Since giving up the possibility of new tour routes adjacent to the cave entrance, the Park Service had encouraged us to find locations in the cave close to the surface where an artificial entrance could be dug. Nearly a mile southeast of the entrance, only a third of a mile south of the Long Winded Passage, the surface terrain is cut by another valley, Lithograph Canyon, where small cave openings may be seen in the limestone cliffs. Everyone agreed this was a likely location for the new entrance.

So we pushed south, crossing the large inviting east-west passages, leaving them for the future. Even so we had surveyed ten miles of cave by the time we reached the Lithograph area.

According to our survey we were only forty feet below the surface where we came closest to the canyon. Not only did this make an artificial entrance seem feasible, but there were interesting and beautiful things to see. Water seepage from the canyon had produced a covering of flowstone and stalactites on the crystal walls. There were slender soda-straw stalactites several feet in length, colorful draperies, and one strip of cave bacon twenty feet long. Moreover, the passages were large enough to accommodate crowds of visitors. Eagerly we mapped a tour loop.

This is almost the exact tour that the National Park Service opened to the public just ten years later, in May 1972. Events moved haltingly at first, with long waits for Park Service and congressional approval, for appropriation of funds, for engineers to come up with plans. Then our survey had to be

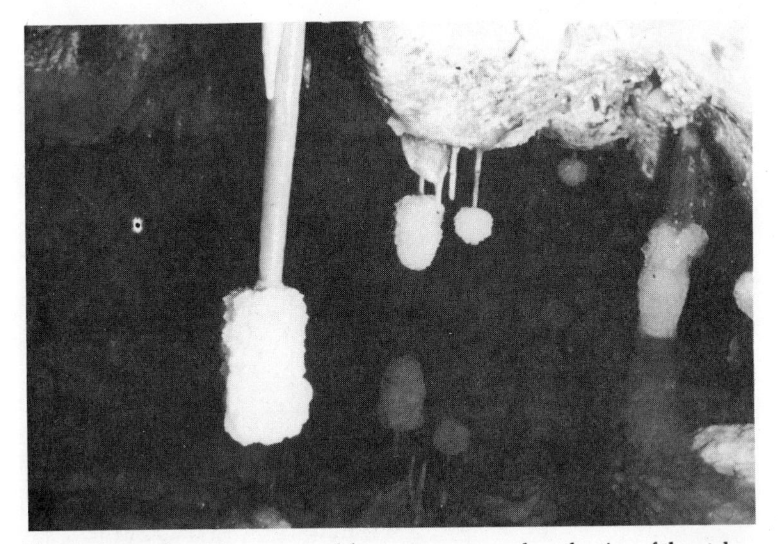

The Bottle Brushes. These unusual formations grew when the tips of the stalactites were partially immersed in a pool. *Dave Schnute.*

verified. No one was foolish enough to propose digging into the hillside at a spot based on a shaky succession of over 200 compass sightings and tape measurements within the cave.

One room on the proposed tour was bigger than the rest and at a higher level. It soon acquired the name "Target Room," since it was proposed for a well driller to try to hit it from the surface. First, however, Father Paul Wightman and Earl Biffle, developers of a cave radio location system, were asked to bring their equipment to the cave. With transmitter and loop antenna broadcasting from the Target Room, the signal was picked up on the hillside above and a spot located exactly above the center of the cave room. Here, to be doubly sure, well driller Bob Nichols was instructed to drill. In June 1964 a six-inch well hole broke through the ceiling of the Target Room. We cheered—but alas, all the cavers we knew were bigger than six inches. The hole was no help to the exploration.

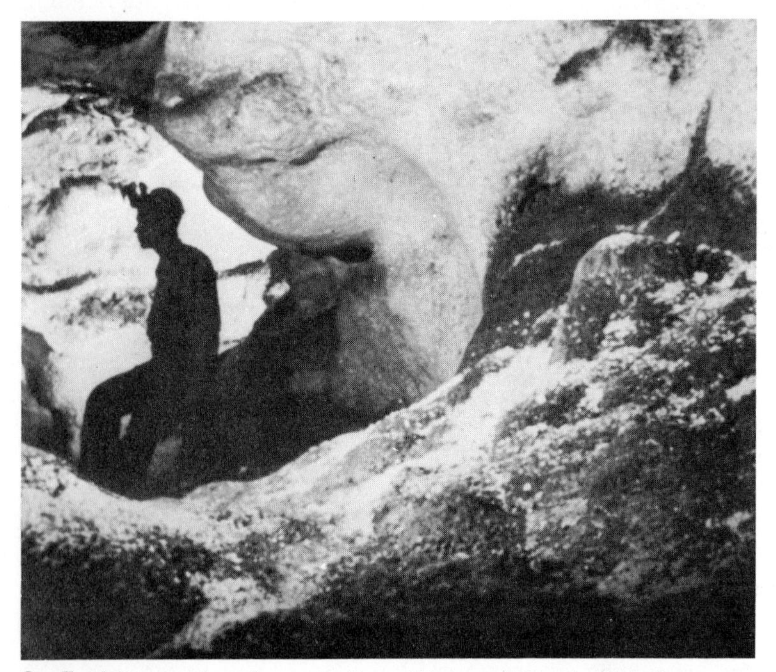

Small calcite crystals and gypsum "snow" festoon the walls. *Herb Conn.*

Furthermore, exploration had now been halted by official order. A few of the people whom we had taken through the Badger Hole, the Long Winded Passage, and all the other trails of the route to see the splendor of the new tour had been unfavorably impressed by the difficulty and danger of the trip. For safety's sake, they said, and for protection of the fragile decorations along the proposed new trail, let's wait until the new entrance is installed.

It was frustrating to wait, knowing about all those unexplored leads in there waiting. But the Target Room tunnel, when it was finished in 1966, saved so much time that it was worth it. The elevator, finally completed in 1972, was to our cave exploring crew sheer luxury.

In addition, visitors to the cave now have a choice of tours. The scenic or modern tour enters the cave through the elevator, and, with the help of modern electric lighting and walkways, passes through some of the prettiest rooms and formations in the cave. The primitive or historic tour goes in the original entrance, using the steep rough trails little improved since the days of the Michauds. Here visitors carry lanterns supplied by the National Park Service. A third tour, added in 1974, is the "spelunker tour," limited to active individuals who want to see some of the cave beyond the developed trails. Participants wear hard hats and electric headlamps and spend several hours scrambling, crawling, and sightseeing in the company of a specially qualified ranger-guide.

With the new entrance a reality, mysteries that had been painful hours from the surface were now close at hand. One such was Hurricane Corner, a hole southeast of the Target Room, where the wind often roared with unnerving force. It had been discovered in 1962 by Alan Howard, speleologist and former guide at Wind Cave. With him we had squeezed through this windy aperture and explored some of the cave beyond, debating as we did so why the wind was so strong here so far from the entrance. We found that here, as at every windy spot in the cave, the wind might be blowing in either direction, or hardly at all. It depended, people said, upon the barometric pressure.

The cave, essentially, is a huge jug with a narrow mouth.

We've all heard the swoosh of air rushing in or out of a jug when the lid is removed and the air within is at a different pressure. Since atmospheric pressure is changing almost constantly, the cave reacts by blowing one way or the other in an unceasing attempt to equalize the pressure within.

The strong wind at Hurricane Corner seemed to indicate, possibly, that another natural entrance was close. We knew that the bed of Lithograph Canyon was no more than fifty feet above it. A hole, even though a tiny one, might be draining the cave here just as the natural entrance did in Hell Canyon a mile away.

Each time we checked with rangers working at the entrance, however, we learned that the wind direction at Hurricane Corner was wrong to indicate another entrance. When the cave entrance exhaled, the wind was in our face at Hurricane Corner; when it inhaled, air was sucked into the caverns beyond Hur-

Herb Conn squeezes through the tight spot at Hurricane Corner as the wind whistles around him. *Dave Schnute.*

ricane Corner. This small blowing hole must lead, not to the surface, but to a vast amount of further cave.

At least so it seemed to us. Another theory, the old one that what blows in one entrance must blow out another, soon presented itself. Air blowing in from Hell Canyon might be blowing out in Lithograph.

At that time it seemed that most cave weather researchers—speleometeorologists you might call them—looked upon barometric winds with a bit of skepticism. Most caves are just not large enough, they felt, to produce a strong wind in this manner. The predominant cause of most cave winds was thought to be the "chimney effect," similar to the draft produced by a hot stove or fireplace which draws air in through doors and windows and blows it out through the chimney. In winter the cave is a warm stove. In summer, colder than the outdoors, the cave produces a "reverse chimney effect," drawing air in from an upper entrance and out through a lower one.

At Jewel and Wind caves, however, the wind direction bears no relation to the season or to the outside temperature. Sometimes these caves blow in one direction for five days or more; sometimes they change direction several times a day.

Curious to learn what these cave winds really mean, we did some research during the period when we weren't allowed in the cave. We built a crude recording anemometer, a large paddle which blew back and forth in the wind, moving a pen which marked upon a rotating drum. With this device and a recording barometer at the entrances to both Wind and Jewel caves we were able to prove that the wind in both caves is a direct result of changing air pressure, a one-entrance wind which does *not* travel to or from a second opening to the surface. If the cave inhales steadily for five days, as it sometimes does, at a velocity averaging 10 miles per hour, 100 million cubic feet of air enter the cave, and all of it remains inside until the wind reverses. (Any other entrances the cave may have are also inhaling.) Obviously the air pressure in the cave must increase with all this added air, but it can't increase any more than does the outside pressure which blows the air in. If 100 million cubic feet of air raise the cave air pressure no more than half an inch of mercury (the rise

The rare mineral hydromagnesite hangs from the walls in semi-liquid balloon-like sacks about 3 inches long. It is sometimes called "moon milk." *Dave Schnute.*

A foot-long "beard" of hairlike gypsum waves gently with the slightest air movement. *South Dakota Highway Department.*

shown on our barometer during one set of readings at Jewel Cave), we figure there must be a total cave volume of at least five billion cubic feet. Herb published these studies in the April, 1966, *Bulletin* of the National Speleological Society.

It's difficult to estimate the volume of the cave already discovered, but one mile of average-sized passage is perhaps equivalent to a million cubic feet. Five *billion* would fill one enormous room encompassing every known passage in the cave, east to west, north to south, top to bottom. Or, spread out in the fashion of the passages already discovered, it would ramble on, mile after mile, under the hills and canyons of western South Dakota—as it probably does.

Following the wind's devious course within the cave is not as easy as it may seem. On our first excursion through Hurricane Corner in 1962, we had been halted just beyond in the Pool Room for want of a rope to descend a fifteen-foot climb. Returning better equipped next time, we climbed into the Cloudy Sky Room (named for broken crystal patches on the ceiling), where the wind seemed to desert us. Slowly circling the room's walls didn't help, for the wind—finally located—was sifting up through cracks between large rocks on the floor. After moving enough rock to allow passage, we were off on an obstacle course headed for the south side of Lithograph Canyon. That trip ended on a discouraging note when the wind divided into a number of cracks, each too tight for us to follow.

With Al Denny, an employee at the cave who had shared in the discovery of the Target Room, we pried out a crystal slab to open up one of the wind cracks. A crawl and a tight squeeze upward brought us into a small stalactite room and easier going beyond. South of the canyon at last, we looked forward to finding another area of big rooms and passages like that beyond the Long Winded Passage.

Briefly things looked encouraging. In what we now call the Delicate Arch Room, the ceiling was 50 feet high and the far end of the room was beyond the range of our lights. The only difficulty was that the intriguing blackness of the continuing passage was out of reach—twenty feet up the wall. Some ingenuity would be needed to get there. At the top of the climb was a boulder we thought we might lasso with a rope.

This attempt, however, waited several years until the tunnel was finished. Earl Biffle and Father Wightman were with us when we carefully maneuvered first a light line and then a full-sized climbing rope over the boulder twenty feet above our heads. It wasn't until we had tested the rope cautiously with our weight and then climbed it, using mechanical ascenders, that we discovered that the boulder, looking so solid from below, was a frail bridge of tumbled crystal slabs. A "Delicate Arch," Earl said.

Worse, the passage stopped 150 feet ahead. Above the climb there was no trace of wind. Somewhere in the Delicate Arch Room, quite possibly somewhere up under that lofty fifty-foot ceiling, the wind had escaped us.

During the next few years we explored other passages, filling in various new areas of the map. The survey crept gradually upward to twenty-five miles. Each in its own hiding place, we encountered an assortment of curiosities—hollow stalagmites up to twelve feet tall, twisting fingers of chert covered with sparkling quartz crystals, waving white beards of hairlike gypsum crystals, and frail silvery balloon-like sacks of the rare mineral hydromagnesite.

In December 1968 we were on the trail of another wind. While the Hurricane Corner wind seemed to lead southeast, this one was going west. With Dave Schnute we chased it on hands and knees through jumbled piles of rocks we named the Tenderizers, to a tighter spot where the wind hummed audibly and filled our eyes with fine white particles of gypsum. Beyond this Snow Blower we climbed into a loft and down the other side. After we had forced a passage through the Small Favor, thankful it was no smaller, we worked our way steadily downward—still heading west—until the survey showed we were crossing under Hell Canyon.

For a few minutes we thought we had entered an extensive new area of the cave west of Hell Canyon. Hell's Half Acre is a nice big passage decorated with stalactites and draperies resulting from the canyon drainage. Dripping water collects in curious cuplike growths which cavers call "bird baths." The passage, we felt, would lead us on and on. But no, it stopped dead a few hundred feet west of the canyon.

We returned, searching the walls of the room for the opening the wind must have taken. It fairly screamed at us as we passed the Horn, where the wind blows harder than through Hurricane Corner. We clocked it one day at thirty-two miles per hour. Such a wind not only extinguishes your light; it removes your hard hat and sends it bounding on ahead of you.

On the map the Horn looks a bit like Cape Horn at the southern tip of South America. The wind and the route into the cave reverse direction sharply, returning toward the east. Not far ahead, at High Water, we had to battle wind and water at the same time. The wind squeezed across the surface of a small pool, kicking up waves that suggested a rushing river. With a limited amount of air space about the water, it's difficult for the caver to keep dry when he wriggles through, and with the chill factor produced by a thirty-mile wind, a wetting isn't attractive. We later learned to siphon water from this pool, thus lowering its level and making a dry passage easier.

The regions 'round the Horn, past Hell and High Water, kept us busily mapping for several years. Some of this far cave comes very close to the new tour area near Lithograph Canyon, but no shortcut has ever been found. At the Plug Ugly, air seeps through tiny cracks connecting the two cave areas, and a shouted voice can be heard from one side to the other.

The wind threads its way through the complexities of this area, leading mainly south into the *V* where Lithograph and Hell canyons join. Amid a confused jumble of immense fallen blocks, all coated with a slimy layer of mud, we lost the wind for a time. It turned up at last, having ducked into an inconspicuous hole to the east through which it crossed under Lithograph Canyon.

To save carrying drinking water from the pools near High Water, we placed a cut-down plastic jug under the drips at this canyon crossing. Sometimes it will fill in the week between our trips, sometimes not. We cross our fingers as we begin our trip, hoping the jug will have filled, but *not* the pool at High Water.

East of the canyon the passage swells to an expressway, continuing for a quarter of a mile to a point just 300 feet south of the Delicate Arch Room. After learning this interesting fact from our survey data, we returned, hunting for winds in either direction. Surely the Hurricane Corner wind from the Delicate Arch Room joined our new passage somewhere along there.

But our luck was bad that day. All the open leads on which we had pinned our hopes quickly stopped. Although the wind was still blowing steadily along the way we had come, it vanished in the space ahead.

Eventually we traced it to a hole in the ceiling. In a parallel passage to the south we continued east another 500 feet. Here the passage climbed to the upper levels and pinched out. Clearly some further detective work, and perhaps some digging, would be needed to find the way on.

But before making many more trips around the arduous and time-consuming circuit of the Horn, we thought it worthwhile to search again from the Delicate Arch Room side. Wind *must* be using a shortcut across, though we hadn't found its route from either side.

With the Delicate Arch Room now a mere thirty minutes from the new entrance, it was feasible to take all sorts of climbing gear that far into the cave. In the summer of 1969 we engineered a climb all the way to the ceiling above the Delicate Arch, but still found no wind. The mystery was partially solved later when we found air sifting from red dirt at the east end of this room. After burrowing like moles for sixty feet through this dirt and assorted small rocks, we emerged at last into a passage. Somewhere we picked up the main current of the wind again. It reappeared as mysteriously as it had vanished, ready to greet us at the Humdinger with a deep-throated roar. Beyond are innumerable crawlways, drafty and unpleasant, known familiarly as the Miseries.

Finally every winding arm of the Miseries had been explored as far as it was possible for us to squirm. Still there was no connection to the cave to the south, nor was there a route beyond. Wind teased us through five or six openings where we couldn't follow. We turned back in defeat, ready to return to the Horn and High Water.

In the summer of 1973 after a couple of rainy years, High

Water was especially high. Keeping the pool drained enough for a dry crossing proved not worth the effort.

Two girls of miniature dimensions, Sandy Ramp and Deon Simon, were working at the cave. Both took to caving and soon changed our ideas of what size opening a human being can squeeze through.

With these two mini-cavers we returned to the Miseries, to the tight holes at the far end where the wind got away from us. The first hole proved too tight even for the girls, but the second—a mere four inches high when we found it—could be scooped out to about six inches, allowing the little people to slip through. While Sandy called back words of encouragement: "It goes! There's a strong breeze!" we continued enlarging the Calorie Counter until we could heave our own 7½ inches of bone and scar tissue through the trouble spot.

After 700 feet of Mini-Miseries we stared off at last into Metrecal Cavern, the beginning of much bigger cave passages and a series of switchbacks which dovetail neatly with the far area reached from the Horn. Although we suspect we have picked up the Horn wind somewhere along the way, ironically we still can't make the connection.

A strong wind through the 1,500-foot length of the Mind Blower led us farther east than we had ever been, one and a half miles in a straight line from the cave entrance, past spectacular gypsum deposits, on to who knows what. In this region the survey reached a fifty-mile total in December 1973. Ranger-friend Dennis Knuckles, to celebrate the occasion, removed his red shoelace and stretched it across the passage (fortunately it was a long shoelace and a narrow passage) for us to sever with our lamp flame. Besides this ribbon-cutting ceremony, the new boss at the cave, Steve Hurd, and his wife Judy greeted us with pink champagne when we surfaced.

The fact that the cave wind still blows at our farthermost point tends to bear out the theory that the first fifty miles are only the beginning. It would be pleasant to give up (or give out) while there are still open leads to check away out there. But as long as we haven't finished surveying Jewel Cave, we won't feel any need to look for a bigger one.

Fern—Cave of the Pits

Originally all that was known of Fern Cave in Alabama was Surprise Pit, a vertical shaft more than 400 feet deep. Cavers came to it from all over to sharpen, practice, and perfect their vertical caving skills. Surprise Pit still serves as a testing ground for new techniques and equipment for safe and fast ascending and descending.

As more of Fern was discovered it was realized that this was a major cave system, and that Surprise Pit, unique as it was, was just a small part of it. Slowly an intertwining maze of passages and rooms was revealed. Many additional vertical shafts were discovered, plus large rooms, beautiful formations, and Indian and animal remains; there is still more cave to be explored.

Donal Myrick has been one of the chief explorers of Fern Cave; his patient work weekend after weekend has helped reveal its mysteries.

Fern—Cave of the Pits

DONAL R. MYRICK

Every caver dreams of discovering and exploring the perfect cave. But what is the perfect cave? I believe it must have something to offer every breed of spelunker. It will provide adventure for everyone. It will provide challenges for the hardy explorer, yet have some passages easily accessible to the average caver. It will possess a variety of large passages, small passages, tight squeezes, giant rooms, waterfalls, beautiful formations, delicate mineral crystals, deep dark pits, and—most important—it will have potential for discovery of new miles of unexplored passages and pits.

From my point of view the perfect cave was discovered June 4, 1961, when members of the Huntsville Grotto of the National Speleological Society visited a sinkhole high on a mountain near Huntsville, Alabama. This was the entrance to Fern Cave.

The leaders of this first expedition to Fern Cave were Jim Johnston and Bill Torode, both of whom played major roles in the exploration and mapping of the gigantic system. Fern Sink, named for its profusion of ferns, has two cave entrances, one wet and one dry. Water cascades thirty feet into one entrance. Everyone but Bill went to look in the waterfall entrance.

A giant pillar in one of Fern Cave's many large rooms. *Alex Sproul.*

Bill thought they were going into the cave, so he took a quick look at the waterfall and then went into the dry entrance. But Jim and the rest had sat down outside to eat lunch. Bill scurried down the passage thinking the group was ahead of him. After 600 feet he came to what is now called the Waiting Room. He kneeled down at the edge of a waterfall which disappeared into the dense darkness below. Bill at that moment became the first caver, and perhaps the first person, ever to peer into the depths of Surprise Pit.

The standard method of estimating the depth of a pit is to toss in a rock and count the seconds until you hear it hit bottom. Bill threw in a rock and started counting. He did not hear it hit, and assumed that the noise of the waterfall had muffled the sound. He tossed in a larger rock. Several seconds passed before he heard the distant muffled reverberation. He went back to the entrance and brought the rest of the group. More rocks were tossed in and it was concluded that it was a four- or five-second pit, about 350 feet deep. Since they didn't have enough rope for a pit of such depth all they could do was look.

In 1961 the deepest known pit in the United States went down 350 feet. Initial estimates of Fern's depths indicated that a new record might well be set. No one in the Huntsville Grotto had any experience descending pits like this; in fact, no one had any ropes long enough for the task. So we called on the expertise of Bill Cuddington, a pioneer in the field of vertical caving.

Bill brought a 368-foot length of half-inch manila rope to use as the main rappel and line. This was rigged directly to an expansion bolt driven into the solid rock, and backed up by two pitons (metal spikes) anchored in small cracks. An electric light was attached to the rope's end, and it was lowered into the pit. When all the rope was lowered the light was seen spinning far below on the end of the rope. The rope was pulled back up, and the light was tied to a safety line. This was then tied to the original rope, and again lowered into the pit. This time the rope reached the bottom. The rope was later measured, and the depth of Surprise Pit was determined to be 426 feet. Years later more accurate methods found the depth to be 437 feet, or almost a six-second pit.

Bill donned his sling set and rappel spool, hooked onto the line, and began his record-breaking rappel into Surprise Pit. Thirty-five minutes later, he reached the bottom of the then deepest pit in the United States. After a brief look he began the long climb back out, using foot slings attached to the rope. One hour and thirty-one minutes later he climbed over the lip at the top of the pit.

Excitedly Bill told the group, "That's a lot of hole! You're against the wall for the first ninety feet or so and then it bells out. It's a free fall all the way. It tops anything I've ever seen!"

The fame of Surprise Pit quickly spread, and cavers came to Alabama with the sole purpose of descending the pit. But this caused problems; the majority of visiting cavers who came to Fern were not adequately prepared, having neither the equipment nor the experience. Gradually the challenge presented by Surprise Pit caused a revolution in vertical caving techniques and equipment. Numerous rappel devices were conceived, built, and tested solely to do Surprise Pit. Today 1,000-foot rappels are almost routine. In 1961 Bill took one hour and thirty-one minutes to climb out. Now with metal mechanical ascenders the record for Surprise Pit is well under twenty minutes, and most cavers can do it safely under forty.

For many years Surprise Pit was all that was known about Fern Cave. Almost seven years later we received a report of a seven-second pit in the same mountain near Fern. The existence of a seven-second pit in Alabama was hard to believe, since this would imply a depth of more than 650 feet. However, a six-second pit—about 450 feet—was indeed possible, and we were all tremendously excited. A few weeks later we managed an expedition to explore this new pit. We lugged an 850-foot rope up the mountain and searched around and around the reported locale until we found a gigantic sinkhole.

The sink contained two pits in the bottom. But rocks echoed back in less than three seconds. Well, it wasn't a seven-second pit. There we were with our 850-foot rope primed to do a 700-foot pit, but the deepest of these two pits was only 105 feet. The north pit was a dead end. The south pit led off into some interesting

A caver rappels into the depths of Fern Cave. *National Speleological Society.*

but eerie passages past several recently dead bats in various stages of decay.

We made three trips in succession to this cave, and a great deal of passage was discovered and explored. Most of the cavers who saw the dead bats resolved never again to drink cave water. Perhaps it was this queasy feeling we all experienced that caused us to drop the exploratory efforts in this cave even though it provided the most promising leads for the continued exploration of Fern in over seven years. For obvious reasons we named this cave the Morgue.

Bill Torode began mapping the top of Surprise Pit, where inaccessible passages could be seen leading off into the darkness, by digging a ledge in a muddy layer around the top. He finally made it to the far side and set out solo, exploring about 900 feet of passage.

Spurred by Bill's success, exploratory and mapping efforts in Fern began to increase. During a scouting trip Jim Johnston and Bill Torode were checking for possible entrances. North of the cave they noticed what appeared to be a large washed-out sinkhole. They found a small crack with air blowing strongly out of it. It didn't look like much, but Jim squeezed in.

This discovery, an unimpressive little hole, led into the most extensive cave system in the southeastern United States. The most surprising thing was that it took almost eight years. The entrance is only a thousand feet from the Fern entrance, yet the hundreds of cavers who visited Fern each year failed to find it. And we were not the first to enter. In spite of the smallness of the entrance, some hardy soul crawled in there in 1861 and 1864, and recorded his visits in the mud. From these graffiti we know that some of the hill people were aware of Fern's entrances at least back to Civil War times.

Bill had left his caving pack and light back at the Fern entrance. Jim had his light, so he hurried down the virgin passage with the hopes of discovering another 400-foot pit or perhaps another approach to Surprise Pit. As he went he noted many side passages and the decision as to which to explore first was difficult. He was highly excited, since he expected each shadow and depression in the floor to drop off into some bottomless abyss. Near the entrance was a large pit which echoed loudly when rocks were tossed in. It was not, however, the expected 400-foot-deep chasm.

One of the passages was decorated with delicate helictites and terminated in a thirty-foot pit. Jim returned to tell Bill about his find, and together they hurried back to tell the others.

This new entrance was named the Johnston Entrance, and because everyone felt that the new cave was merely part of the well-known old Fern, the cave was called New Fern. We thought the two caves would quickly be joined. The next day we had two groups in the cave, one group led by Jim exploring the new passage, and the other by Bill Torode mapping. Everyone wanted to be present when the connection to Fern was made.

Bill began by mapping the most obvious passages. He went in virtually a straight line for 3,500 feet directly toward the Morgue Cave. This, named the West Passage, was seen to pass within eighty feet and slightly above Surprise Pit when plotted on the map. This observation caused many hours to be spent on numerous trips combing the West Passage for the elusive connection to Fern. On one of these trips Bill Torode and Dick Graham connected New Fern with the Morgue; later a connection was established between the back room in the Morgue and some pits in the West Passage. However, the connection with Fern was not found.

The other group went into another passage which veered off to the east more deeply into the mountain. It turned out to be a multilevel maze crisscrossing itself in many places. We named it the East Passage. The deep pit near the entrance proved to be only 117 feet deep, with nothing but tight crawlway passages leading out from it.

We went down the long, narrow canyon passage decorated with delicate helictite formations. This passage terminated about 2,000 feet from the entrance in a thirty-foot pit. While the pit was not deep, it was difficult just to get a rope back to this point.

Above the passage was another level with a series of large and beautifully decorated rooms. These Formation Rooms are easily accessible and have since become the most popular rooms in the entire cave, photographed by hundreds of cavers each year. As a consequence they have suffered the most damage of any area. The passage continues for several thousand feet and eventually connects with the West Passage. Not too far beyond the Formation Rooms is a thirty-foot pit, called the Blowing Hole for the noticeable flow of air coming from it. Bill Torode was the first to notice this. He cleaned off the area around the top of it, pushing a sizable amount of dirt and rocks down the pit. He was later to find this pile of dirt by coming into the same room from a passage at the bottom.

Trips to New Fern were an every weekend-all weekend affair. It was not unusual to have two or three groups in the cave simultaneously. But the more we explored the cave, the more the connection of Fern and New Fern seemed to be only a dream. The three miles of mapped length in New Fern now made it the second longest cave in Alabama. Our primary objective began to shift from connecting with Fern to exploring and extending New

Helictites adorn a flowstone wall near the Formation Rooms. *Jim Johnston.*

Fern. Since we already had seen so much new passage, and since we had not come to the end of any of our primary leads, we were very hopeful of making New Fern the longest cave in the state.

A trip was made to explore the passage at the bottom of a deep pit Bill Torode had found. It was located just before the crawlway that leads to the Formation Rooms. We had been within twenty feet of this pit on our first day in the cave, but it is hidden beneath some slabs of breakdown on the way to the Formation Rooms.

The pit was a beautiful 155-foot drop that echoed loudly when we dropped in a rock or shouted. The passage at the bottom led into a meandering canyon that connected with a sizable stream. Downstream the passage began to narrow, the walls became higher and steeper, and the gradient of the stream increased. We came upon a beautiful cascade and yet another

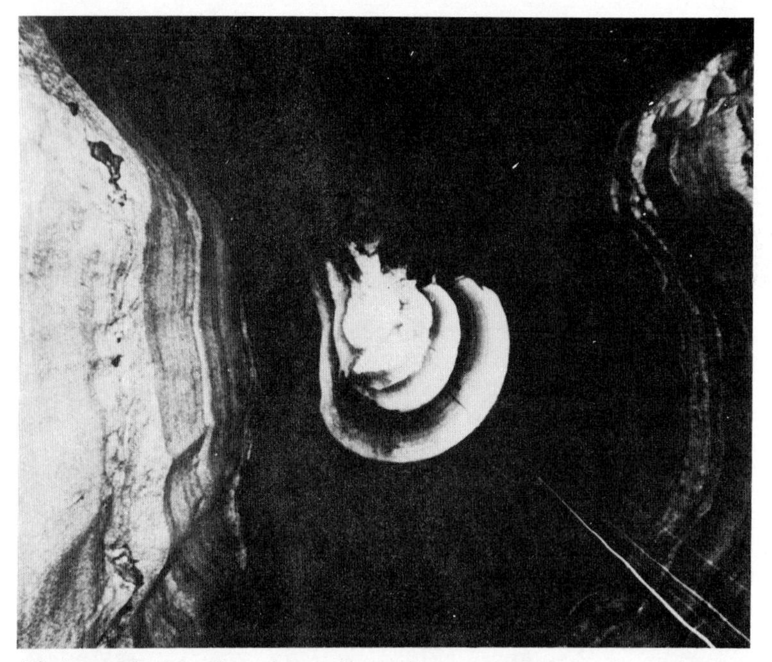

A look straight down from the top of a 200-foot pit in New Fern. The rope used by the climbers can be seen hanging in three loops on the right. *Alex Sproul.*

pit. Although we did not at this time have another rope, I knew we would be back with one.

We regrouped above the cascade and chimmeyed up the narrow canyon, pausing to eat in a little room we named the Lunch Room. Bill climbed through a crack in the ceiling. At the top he found a large passage, a big pile of dirt, and what looked like good passage above a small hole in the ceiling thirty feet above him. This was difficult to climb to, so after he got there he wrote *ENDS* on a small flat stone, and propped it up so that it could be seen from the big passage below.

When Bill plotted the results on the master map, it became rather obvious that he had climbed up to the room at the bottom of the Blowing Hole, and that the pile of dirt was the one he had made when he cleared off the ledges at the top of the pit. This indicated that the water passage and pit we had just discovered were accessible from the Blowing Hole. If so, this would shorten the trip by several hours.

Our map of the cave was growing. Bill Torode did most of the drafting. We quickly learned that if conventional drafting techniques were used, the result would be unintelligible, because there were so many intertwining and intersecting levels and passages. So Bill devised a color scheme to sort out all the levels and criss-crossing passages. Many features in the cave derived their names from their colors on the map, such as the Gold Passage. The master map is over nine feet long and five feet wide.

On the next trip we tried to tie up that pit and the water passage. We entered the cave at the Johnston Entrance, and quickly made our way to the Blowing Hole. We rigged the drop and I rappelled down. At the bottom, I noted with amusement Bill's *ENDS* sign. We later named this the Gold Passage for the color assigned to it on the master map. We went through the crack in the floor to the Lunch Room. That crack is the most complicated maze I know. There are at least a dozen passages going almost from the top to the bottom, and vice versa. All of them involve tight vertical squeezes and difficult chimmeying.

From the Lunch Room we went by the water passage, the cascades, and the ledge at the top of the pit to the dry area I had

Bill Torode explains some of the complexities of the Fern Cave map. *Jack Ray.*

located previously. I rappelled this 48-foot drop and found myself in a large passage leading south.

The passage was beautifully decorated with a stream, and its high, intricately sculptured walls were covered with pure white flowstone with an occasional touch of red and black. After 700 feet the water fell over an eight-foot drop, and cascaded to an even lower level. Fortunately we had a spare rope and rappelled into the room. Later we piled up some rocks to make the eight-foot drop climbable without the rope.

The room was the first of a chain of five huge rooms, by far the largest we had yet discovered. The passage ranged from twenty to sixty feet wide, and was twenty feet high; the blackness stretched farther than our lights could penetrate. We were tremendously excited with our find. We wanted to run down this new passage, but the floors of all the rooms were covered with massive heaps of breakdown, making our progress slow and laborious.

At the end of the fourth big room a high, narrow passage veered off at the ceiling level, terminating after about 200 feet. In it we discovered the first of New Fern's vast and varied gypsum formations. Everything seemed coated with shimmering gypsum crystals, and the walls and floor sparkled in the light from our lamps.

In the fifth big room the large passage seemed to terminate, although a narrow passage continued at the opposite end at ceiling level. We followed this several hundred feet until we found a pit, which we estimated to be about 100 feet deep, but which measured to be eighty. We named this, the sixth big room, the Balcony Room. Continuing on down the narrow passage we came to another large room, but impending exhaustion compelled us to turn around. We knew we would be coming back next weekend, so we left all of the drops rigged with ropes to make it easier on the next trip. This became standard practice, enabling us to push the exploration as rapidly as we did. On almost every trip we discovered a new pit that required the extra rope we had learned to bring along "just in case."

When we pushed on through the seventh big room and traversed a short wide crawl, we found ourselves staring out into yet more impenetrable darkness. This new room was indeed impressive: over 200 feet long, 100 feet wide, and in places 100 feet high—over two-thirds the size of a football field. The only way out appeared to be a 55-foot drop, but there was no passage at the bottom. We thought we had finally come to the end of this tremendous sequence of rooms, until someone noticed a small hole about twenty feet from the bottom of the pit. After a difficult climb we found ourselves in an underground rock pile, a collapsed area of the cave in a canyon filled with large rocks. Our job was to find a way through the jumble. We named this the Jericho Passage, hoping that the walls wouldn't come tumbling down. After 200 feet of squeezing between rocks we found ourselves staring into a deep canyon at the junction of another major new passage. Again we had no more rope, so we had to turn around.

On the next trip we carried a 150-foot rope for this drop, plus the 50-foot "just-in-case" rope. We rigged the pit, which was 72 feet deep, and I rappelled down, finding myself in a long narrow canyon. In one direction I was stopped by a 20-foot pit. I went the other way and found another pit but I couldn't see the bottom with my light. I could hardly believe that a pit could exist down here since we were already over 400 feet beneath the surface, but I carefully bellied up to the edge of the pit and flashed my light directly down. The pit was only about thirty feet deep. Dan arrived with the 50-foot rope. While the others descended the 72-foot drop, I rappelled down and found myself coming through the ceiling of a very large room. The passage was forty feet high, thirty to sixty feet wide, and about 1,000 feet long. It had a sandy floor, a flat ceiling, and even a little brook off to one side. This is what cavers dream of discovering—a new cave. We were all exuberant about this totally unexpected find. We had all thought we would be coming to the end of one of New Fern's passages, but instead we made a major new discovery.

We began surveying, and mapped the main passage and parts of two large side passages. By this time we had been in the cave for over fourteen hours and had at least a four-hour return trip facing us. So, once again, we reluctantly started back out without coming to an end of the passage.

While the others were going up the thirty-foot drop, Dan

Hale and I decided to have a look upstream. It was easy walking. We saw something on the floor that looked liked charred wood. Dan and I examined it carefully. It appeared to be an Indian cane reed torch, burnt ends and all. For a moment we looked at each other in amazement. The Indians had no ropes, and obviously did not go up and down the pits. We must be close to an entrance! We thought we were about to find a lower entrance and would be outside soon. We moved quickly up that passage. We went nearly half a mile and the ceiling became lower. Reluctantly we turned back and caught up with the rest of the group. We told them what we had found and we were all disappointed. Finally, after spending over twenty hours underground, we left through the Johnston Entrance. When we plotted this section, which we called Bottom Cave on the master map, one of the large side passages was seen to be heading directly toward the bottom of Surprise Pit! This was the first time we had found a passage which showed any real promise of connecting with Fern. But we never did find another entrance.

Although Bottom Cave became well known primarily through the merits of its passages and their delights and difficulties, the paleontological discoveries in this area have also excited interest. We found well-preserved bones of several extinct vertebrates, including the red wolf (*Canis rufus floridenis*), the Pleistocene bear (*Arctodus simus*), and a Pleistocene jaguar (*Panthera onca augusta*). It is obvious that at one time an easily accessible entrance to Bottom Cave existed, apparently filled in or covered over subsequently.

Just to get to and from Bottom Cave required a tremendous amount of time and energy, so we made plans for a weekend camp-in-the-cave trip. Prior to that trip Bill Torode led a group to the West Passage to an unentered 200-foot pit directly above the West Room. We hoped this pit would be a shortcut to the Bottom Cave. It came down in the middle of a 200-foot passage, with 70-foot drops at either end. Bill pulled the excess rope over to the southeast drop, but it was too short to reach bottom. He decided to rappel to the end of the rope anyway to see if he could identify the bottom. At the end of the rope he was still thirty to forty feet from the bottom, and he found himself suspended at one end of a large room with a waterfall at the other end. There were footprints on the ledges around the sides of the room. Suddenly he knew where he was; the room was the West Room and we had our shortcut. This route, through Morgue Sink, the West Passage, and the 200-foot pit, would save hours getting in and out of Bottom Cave. The passage meandered back and forth over the top of a deep canyon, and was thickly encrusted with gummy, slick bat guano. For our overnight trip it took two hours to ferry our equipment the 600 feet down that passage to the 200-foot pit.

We arrived at 3:00 A.M. and set up camp in a large dry sandy-floored room centrally located for exploring the Bottom Cave. Our main objective for the weekend was to finish mapping the passage that was heading directly towards Fern, and—hopefully—make the long-sought-after connection. But the passage terminated only a few hundred feet from where I had stopped my survey. Reluctantly we gave up, naming this large room the Disappointment Room. This scratched our hope of connecting Fern with New Fern. But we had spent forty-three hours underground, mapping nearly 7,000 feet of new passage. The mapped length of New Fern was now over thirteen miles, making it the longest cave in Alabama, and, at that time, in the southeast.

Later, hoping to find a new short-cut route to the Bottom Cave via a newly discovered passage in the Morgue, we entered Morgue Sink, descended the 105-foot drop, and, after some confusion, found the desired passage. It did indeed lead to the Bottom Cave, saving at least four hours, and it involved only 150 feet of rope work instead of nearly 400. We also toured the Disappointment Room, re-entering one of the supposedly blind leads going out of the room. It was a belly crawl but opened up again into walking height. We mapped 1,500 feet of new cave. It didn't connect with Fern, but when I plotted the passage, I saw that we had passed within 300 feet of Surprise Pit. I knew then we were on the right track. We had Surprise Pit surrounded on three sides, and we had discovered a small stream coming from the general direction of the pit. Surely we could find a connection.

The next trip Dick Graham descended Surprise Pit to explore

the stream passage that exited at the bottom. Many people had crawled part way down that stream passage. But it was a very tight belly crawl in about six inches of cold water, and no one until now had been sufficiently motivated to keep going. At one point Dick had to dig a channel through a gravel bar that blocked the way. He made it beyond the bar, but he was alone, cold, his carbide light was getting very low; he had to turn back. I led a later trip into the new section, ending up in a little room that appeared to be the final cul-de-sac. I had noticed that the stream disappeared under a low ledge on one side of the passageway, and looked for a place to squeeze under. I began digging with a trenching tool; Jim Johnston spelled me on this onerous task. We dug furiously at the edge of the water, a pool about six inches deep and twenty feet wide. I wiggled and squirmed on through, and on the other side I dug again. With my chin in the water, and my hard hat in front of me, I trenched my way through the gravel bank. I came to another pool of water and could see the passage continuing. But my light was getting low, and the fifty degree water was causing me to shake so uncontrollably I retreated. Everyone was cold, and no one felt like continuing through the long, low, wet trench I had just dug. We named this area Myrick's Masher, which I still feel is an understatement.

Later we planned an all-out push. Our plan was to have two teams: Dick Graham's party, entering Surprise Pit, would try the stream passage; my group would enter the Morgue, and push on through Myrick's Masher. We set up times at which we would try to communicate.

The Surprise Pit crew moved very rapidly, and Dick arrived at the passage well ahead of us. From the Fern side it is not at all obvious where my crawlway comes through the last pool and out from under the very low ledge. Dick looked and looked without success. He tried to communicate according to agreement, but we were far behind schedule and too far away to hear him. After a while he became discouraged, scrawled the date on the low ceiling with his carbide lamp flame, and retreated. Shortly thereafter we arrived at the Masher, and Jim Johnston began digging to lower the water level in the pools. Sherry Graham crawled out ahead of Jim. She pushed on through the still-deep pool, and, emerging from under the low ledge into a two-foot high passage, she found the place where Dick had scrawled the date. The connection was made! She continued on down the low passage toward Surprise Pit, but having left her pack and carbide behind she ran into difficulty. I crawled in after her, cursing the inadequate excavation job. It had filled in a bit from the previous trip and I had to dig it out again before I could get through. On the other side, with plenty of light and enthusiasm, we quickly made our way to Dick's crawlway, which we named Graham's Grinder.

Again I cursed the inadequate excavation job, because again I had to dig in the wet, sloppy gravel in order to get through. Then I emerged from the stream passage into the bottom of Surprise Pit. With much shouting and glee, we were greeted by some of Dick's crew, who were in the process of climbing out of the pit. It was a fine feeling to be the first person to stand at the bottom of Surprise Pit without first rappelling down. We thus brought to an end the long quest for Fern.

We joined together three major cave systems, and mapped over fifteen miles of passage. We have found passages and formations which are equal to the best to be found anywhere. The cave has yielded ancient animal remains, skeletons, and Indian relics that have delighted archaeologists and paleontologists. With its bats and other fauna, Fern has proved to be a cave of prime biological importance. It has historical significance dating back at least to the Civil War. From a sport caver's point of view, Fern has passages and pits for exploration and discovery which are unexcelled anywhere.

For instance, do you recall the breakdown area near the Formation Rooms above the 155-foot pit—the area that is now so well traveled and which we all passed on our first day in the cave? Recently Bill Torode discovered an obscure little hidden passage that begins there. It goes *under* the West Passage and emerges at the *top* of Surprise Pit! So even today there are discoveries waiting to be made in Fern Cave.

Refugees of the Ice Age

Animals living in caves present many puzzling questions and problems for biologists. Why are some cave animals identical to those living on the surface while others have developed special adaptations for survival underground? Why do some caves and cave areas have many different kinds of animal life while others have few or none? How can animals eat, live, and reproduce in an environment of total darkness, sparseness of food, and no daily and few seasonal weather changes?

The southeastern United States has a rich diversity of cave life that has been studied more extensively than most areas. Some caves have as many as forty different species of blind cave animals. Thomas C. Barr, Jr., has been visiting, exploring, and studying caves since he was seven years old. He has done much work in the Mammoth Cave area which has helped answer some of the questions asked above. A biologist at the University of Kentucky, Dr. Barr is a former president of the National Speleological Society. His story is reprinted from *Natural History* Magazine.

Refugees of the Ice Age

THOMAS C. BARR, Jr.

Because of its inherent simplicity, the life community of a cave holds a special attraction for ecologists. Unlike a tropical rain forest, a meadow, or even a farm pond, a cave is inhabited by a comparatively small number of animal species. In the eastern United States, caves usually contain between ten and twenty blind, cave-limited species. Although some may have as few as five or six, other, more richly diverse cave communities may have as many as forty such species.

Mammoth Cave in Kentucky is the best known of the rich cave communities, and biospeleologists have recorded about forty-five blind species among the more than two hundred species of animals found in the system. One of the best places to study Mammoth Cave's terrestrial fauna is in White Cave, a small cave connected to Mammoth by narrow crevices impenetrable to man and rarely visited except by biospeleologists.

When a visiting Australian colleague wanted to see the famous Mammoth Cave fauna, it was only natural that I took him first to White Cave. As we climbed about, over, and finally through a long curtain of dripstone that slices diagonally across

---

This eyeless millipede feeds mainly on cave cricket guano. *Howard N. Sloane.*

the cave, we saw here and there a small creature crawling rapidly over the wet rock. Farther in, the cave animals performed as if on cue, one species after another parading before the advancing circle of lantern light. My colleague noted that it would take twenty years to see that much life in the caves of Australia. The remark was a chance one, but it returned to puzzle me for months afterward. How does a community of cave animals arise in the first place, and why should there be such marked disparity between the communities of different cave systems?

Whatever the diversity, the fauna within a cave consists of unusual animals scuttling across wet rocks, swimming in dark rivers, or flying through echoing silence. These animals live in an environment without light, isolated from the surface yet dependent upon it, for there must be sufficient access to the surface for the animals to receive food, water, and air. Through an awareness of the behavior of cave animals, their adaptations to the cave environment, and the environment itself, answers to questions concerning cave fauna diversity can be more readily resolved.

Most cave communities include many troglobites, blind species with reduced pigmentation; a number of troglophiles, animals that can live out their lives equally well in caves or in cool, dark places above ground; and a few trogloxenes, animals that regularly utilize caves for shelter but return to the surface to feed. Of the animals recorded in Mammoth Cave, 22 percent are troglobites; 36 percent troglophiles; 22 percent trogloxenes; and 20 percent "accidentals" that fell or were washed into the cave.

Trogloxenes are often indispensable to the energy budget of cave ecosystems. Bats and cave crickets, for example, which feed outside the caves at night, contribute large amounts of guano, and often even their own dead bodies, to the food web of the cave. Eyeless species of carabid beetles in both Texas and Kentucky have independently evolved the habit of feeding on eggs of cave crickets, which they dig out of the silt on cave ledges and floors.

Water, like food, also enters caves from the surface, the amount fluctuating seasonally. It either seeps downward from the soil through open joints in the limestone or flows underground as sinking streams, carrying food for the cave community in the form of bacteria, protozoans, other micro-organisms, twigs, leaves, sticks, and even logs and larger animals.

In the southeastern United States, the driest time of the year underground is autumn. Transpiration of green plants removes vast quantities of soil water during the growing season, and the normal autumn rains do not saturate the soil until early winter. The first heavy rains of mid-December or early January mark the beginning of the flood season in caves. Even in tropical caves there are usually seasonal floods corresponding to the rainy season at the surface. The food supply carried underground by water is thus concentrated during part of the year, rather than being spread out evenly.

Most large caves "breathe," pouring cooler air from their

Speleologist Howard N. Sloane examines hundreds of hibernating Indiana bats, *Myotis sodalis,* clustered on the ceiling in a secluded part of Mammoth Cave. *Bruce Sloane.*

mouths in summer and sucking in cold, dry air from the outside in winter. This breathing effect is primarily the result of a density gradient between heavier, cooler air and lighter, warmer air. Cool, dry winter air is especially hazardous to troglobites, most of which have lost not only eyes and pigments but also the ability to regulate water loss through evaporation from their integuments.

I suspect that my interest in caves and cave life was first kindled on a tour of Mammoth Cave in the late 1930s. A stubbornly curious seven-year-old, I teetered precariously but persistently on a ledge at the end of the commercial trail, trying to penetrate the gloom of the chasm below by the dim, pulsating light of a hissing gasoline lantern. Ignominiously snatched from my vantage point by my horrified parents and the cave guides, I resolved that I would some day explore far beyond that chasm and a half-dozen others that had roused my curiosity. Sixteen years and five hundred caves later, it was no accident that the Mammoth Cave system assumed a role of central importance in my studies on cave invertebrates.

The principal cave area of Mammoth Cave National Park lies beneath three sandstone-capped ridges, which are separated by blind valleys, a feature in karst areas formed from the coalescence of many sinkholes. The older and best-known cave passages are in Mammoth Cave Ridge, but extensive cavern development exists beneath Joppa Ridge to the west and especially beneath Flint Ridge to the east. The rock strata are gently tilted to the northwest toward the 300-foot-deep canyon of the Green River, which meanders across the park from east to west. Ancient subterranean streams have carved out five or six successive levels of passages. The cave rivers carry water from the streamless sinkhole plain outside the park and eventually empty into the Green River as large springs. Between the sinkhole plain and the Green River, these streams flow through the lowest levels of an incredible maze of passages whose total length makes Mammoth Cave the world's longest known cave system.

In the "Xanadu" of Coleridge's poem "Kubla Khan," the sacred river Alph ran "through caverns measureless to man." In the intricate complexities of Mammoth Cave we can appreciate the poetic similarity. For many years the "traditional" length of the cave was 150 miles, but the prosaic objectivity of the measuring tape indicated that as of the late summer of 1972, all known and mapped passages in Mammoth Cave did not exceed a total length of 55 miles, while more than 85 miles had been surveyed in the adjacent Flint Ridge Cave system.

Then, in early September, explorers from the Cave Research Foundation and the National Park Service crawled and waded their way from the Flint Ridge Cave under Houchins Valley and emerged in Mammoth Cave near Echo River, demonstrating a direct link between the two huge systems. The upper reaches of Echo River extend under the edge of Joppa Ridge, but no direct connection penetrable by man has yet been found between Mammoth and the large caves of Joppa Ridge. Exploration is continuing, and it does not seem too unrealistic to predict that the system may some day reach an aggregate length of 250 to 300 miles.

As explorers probed the vastness of the cave system, the

Cave salamanders live in moist regions near cave entrances and frequently feed outside. *Howard N. Sloane.*

biological exploration of Mammoth Cave kept pace. Historically, it was the larger aquatic troglobites that attracted attention. Echo River, discovered some time after 1837, is inhabited by the cavefishes *Amblyopsis spelaea* and *Typhlichthys subterraneus*, the white crayfish *Orconectes pellucidus*, and the blind shrimp *Palaemonias ganteri*.

The upper part of the river, in what is called the "Roaring River Passage," is an excellent spot to observe aquatic troglobites. In one section of the passage, which we have named the "Shrimp Pools," blind shrimp use their spindly, basketlike mouthparts to strain the bottom silt for micro-organisms. Cavefishes also cruise the pools, feeding on the unwary blind isopod *Asellus stygius*, the amphipods *Stygobromus exilis* and *S. vitreus*, and possibly on stray shrimp. Cave crayfish, attracted by the slightest ripple in the water, wander about over the bottoms of the pools, feeding on whatever living prey or dead food they can find.

The most diverse terrestrial assemblage of animals occurs in moist, stalactitic areas. There, dense colonies of the cave cricket *Hadenoecus subterraneus* crowd upside down on the ceilings, blanketing floors and rock shelves beneath with a thin, crumbly layer of black guano. Detritus feeders—millipedes, bristletails, snails, scavenging beetles, and springtails—utilize the cricket guano and are in turn eaten by spiders, mites, pseudoscorpions, harvestmen, and predatory beetles.

Ancestors of most terrestrial troglobites—living in deep soil and humus—were probably preadapted to a cavelike environment. A richly diverse, moist forest cover seems to favor development of a soil fauna with many species preadapted for cave life; the sparser the forest and the fewer species of trees it contains, the less varied should be the troglobite fauna of nearby caves. This helps to explain the paucity of fauna in Australian caves and my colleague's astonishment at the diversity of Mammoth Cave animal life. The gradual drying of the Australian interior during the Pleistocene probably led to extinction of much of the forest floor fauna, rather than to its survival in caves.

Although forest diversity may provide a partial explanation of the contrast between Australian and Kentucky cave faunas, why should two cave systems in moist, forested regions differ greatly in community complexity? Maximum forest diversity occurs in the tropics, and if this were the sole factor controlling troglobite speciation, we should expect large numbers of troglobites in tropical caves. But this is simply not the case. While tropical caves are occasionally inhabited by aquatic troglobites, terrestrial species are virtually absent.

The richest known troglobite faunas occur instead in regions where Pleistocene climates fluctuated between cool and moist periods during glacial maximums and warm and dry periods during the interglacials. Species ancestral to the troglobites occupied cool, moist forests. They could have become widespread in surface habitats during periods of glaciation, gradually being restricted to forested ravines, sinkholes, and caves as the ice sheets retreated and temperatures rose. For a period of time they could have colonized caves and existed as troglophiles, but with increasing temperatures and decreasing precipitation they would have become extinct in the forest floor,

Guano of the cave cricket, *Hadenoecus subterraneus,* provides food for detritus feeders. The cricket is really a voiceless long-horn grasshopper. *Charles E. Mohr.*

surviving only as cave isolates.

If the ancestors of troglobitic beetles and other arthropods passed through a "precave" stage in the forest floor, might there not be a few such ancestral types still surviving in some cool, moist forest, deep in forest floor leaf litter? It was just this question that took me to the Monongahela National Forest in West Virginia. My feet sank lightly into the thick mat of spruce needles that carpeted the forest floor as far as I could see, except for a sprinkling of mossy boulders. After prying up the root-bound boulders for an hour to look deep into the litter mat, I was ready to admit that I was wasting my time looking for Pleistocene beetles in the twentieth century. Turning over one last boulder, however, I saw a small, red beetle. Waving its antennae uncertainly to and fro, it crawled rapidly across the wet black humus. I was momentarily so absorbed with the thought that I had been transported back to the Pleistocene that I almost missed catching it. It was a very close relative of a troglobitic species that inhabits the Greenbrier Valley caves far below, living proof that ancestors

The northern cavefish, *Typhlichthys subterraneus,* is one of two cavefish species found in Mammoth Cave. *Howard N. Sloane.*

of cave beetles could have been already preadapted to life in cool, moist, dark microhabitats in forest floor litter.

Troglophiles are not genetically isolated merely because they have entered a cave. The same cave may be invaded again and again by individuals of the same species, and any tendency to diverge genetically from the parent species will be swamped by interbreeding between the established population and the new immigrants.

Periodic opening and closing of caves through natural geologic processes is not a plausible hypothesis to explain the isolation that must have occurred for ancestral troglophiles to evolve into troglobites. The smaller cavernicoles, in particular, can readily gain access through innumerable sinkholes and small crevices made by seeping water. If a cave were so tightly closed that even these openings were excluded, the food supply from the surface would also be cut off. Only a moderately large aquatic animal could conceivably be isolated fortuitously in a cave, perhaps being swept over a high underground waterfall in numbers large enough to constitute a founder population, and it is only the aquatic animals that seem to become the infrequent tropical troglobites. The concentration of terrestrial troglobites in regions subjected to glacial climates is strong evidence that alternating Pleistocene climatic regimes provided the isolation that permitted troglophile populations to diverge from their surface ancestors.

Once isolated and unimpeded by swamping, the troglophile colonies could proceed to change genetically and become more closely adapted to their cave environment. Inbreeding in small populations leads to reduction in variability, and natural selection becomes more severe, creating a sort of genetic bottleneck through which incipient species must pass. Most isolated, inbreeding populations of this sort, unable to successfully pass through the bottleneck, become extinct; the few that are successful emerge with a modified set of genetic instructions closely tailored to meet the requirements of their immediate environment.

We would expect that the probability of successful cave colonization and subsequent speciation into a troglobite is rather low, but given a large number of caves and a lot of "trial" populations of troglophiles, we could also expect a "success" now and then.

Some pieces of the cave diversity puzzle finally began to fall into place when I began to compare regional faunas and geographical speciation between the caves of the Appalachian valley in Virginia and east Tennessee with those of the Mississippian plateaus in central Tennessee and Kentucky. Although both areas are underlain by thick strata of cavernous limestone, in much of Virginia, Tennessee, and other eastern states, the limestone had been severely uplifted and buckled with the formation of the Appalachian Mountains. Farther to the west, however, the limestone had been relatively undisturbed. Due to the many faults in the Appalachian valley, the limestone areas of that region had become separated from each other as the rock strata had broken and tumbled over. The softer limestones eroded away and became the floors of long valleys sharply separated by ridges of more resistant sandstone and shales where there were no caves. The undisturbed limestone to the west, however, developed extensive cave systems connected to each other through continuous chains of passages and crevices.

Many elements of the cave fauna were—and still are—incompletely studied, so a total faunal analysis is not possible. I chose to look closely at trechine beetles, small, reddish, eyeless predators widely distributed in caves of eastern United States. A specimen from the back of White Cave had, in fact, been responsible for starting me in the taxonomy of carabid beetles, and when I began this study I had accumulated ample information on more than 150 species of American cave trechines.

Study of cave beetle distribution soon showed that the Appalachian valley had about three times as many species of cave beetles per square mile of limestone terrain as did the broad and cavernous plains of Mississippian limestone to the west. This is an excellent example of geographic speciation.

A cave in the Appalachian valley is connected only to other caves in the same narrow strip of limestone. Exposures of limestone in this area are, with a few exceptions, highly discontinuous, and a specific beetle species is usually found in only one or two caves, the expansion of its range being prohibited by numerous sandstone and shale barriers. Only four of several hundred caves have two species, and none has three or more.

Individual Mississippian plateau caves, however, often have two or more species inhabiting them. A kind of world record for cave beetle sympatry seems to have been set in White Cave, where no less than six species of eyeless trechines coexist. The caves in the continuous, unbroken belts of Mississippian limestone are presumably connected to each other by innumerable cracks and crevices; thus, beetle species have extensive ranges, occurring in many caves over wide areas.

In areas of continuous cavernous limestone similar to the karst of central Kentucky, a troglobite can disperse by moving from cave to cave through underground openings. From this point of view, a cave becomes a sort of window through which man can sample a troglobitic species at different points in its geographic range. Theoretically, the species can appear in every cave community in the same continuous limestone mass, so any given community can be composed of species whose ancestors successfully colonized different caves at different times in the past.

Such communities may also include a few widely distributed subterranean spiders, springtails, and small crustaceans, which in addition to having all the attributes of troglobites, are equally well at home below ground in nonlimestone terrains. And to the troglobites must be added a variety of troglophiles and trogloxenes of recent vintage, all three groups of animals interacting with each other and with the microbial flora and fauna to sustain the dynamics of the cave ecosystem.

These observations on beetle speciation and distribution are best explained by a simple model. Let us imagine two areas underlain by cavernous limestone, one quite large, with one hundred caves; the other small and with only one cave. Supvose each cave is colonized by a troglophile beetle that, because of a warming climatic trend, subsequently became extinct at the surface. As we have seen, the likelihood of making the transition from troglophile to troglobite is rather low. For each cave population there will be a fixed but low probability that the population

The pipistrelle is a solitary bat, rarely found with others. *Charles E. Mohr.*

will successfully evolve into a troglobite—arbitrarily we can assume one chance out of a hundred. In the small area there is only a 1 percent chance that a new troglobite will evolve, but in the large area the probability is one hundred times greater; that is, it is almost certain that at least one troglobite will evolve there. Since the large area is continuous and cavernous and not broken up by shale and sandstone barriers, interconnecting crevices exist between the caves through which the troglobites can crawl. It is only a matter of time before all of the hundred caves are occupied by the troglobitic beetle that originally evolved in only one cave.

If, despite the odds, a troglobite did evolve in the single cave of the small area, its geographic range would be limited to that one cave because it could not crawl out of the cave and move overland to another cave.

If we now take our hypothetical model and suppose that the caves are colonized not only by beetles but also by spiders, millipedes, springtails, bristletails, isopods, amphipods, and so forth, we would have a closer approximation of what actually took place in caves of the eastern United States. Further model refinements are possible, particularly in weighting probabilities of speciation differently for different sorts of ancestors, but in the present simplified form it is adequate to help us understand how cave communities develop.

In applying our model to the Mammoth Cave community, we can analyze the geographic ranges of its forty or more species of troglobites. These include flatworms, snails, spiders, harvestmen, pseudoscorpions, mites, copepods, isopods, amphipods, crayfish, shrimps, springtails, bristletails, beetles, and blind cavefishes. A look at troglobite ranges shows that there are no more than five or six species strictly limited to the system. On the contrary, the troglobites have a decided agglomeration of range overlaps in the park area. For example, Mammoth Cave is simultaneously the northern limit of one cavefish, *Typhlichthys subterraneus* and the southern limit of another, *Amblyopsis spelaea*, and it is the only known cave where the two occur together. The eyeless beetle *Pseudanophthalmus menetriesii* ranges along the narrow Pennyroyal plateau both north and southwest from Mammoth Cave, but its close relative *P. striatus* extends far out into the sinkhole plain caves southeast of the park, coexisting with *P. menetriesii* only along the inner edge of the Pennyroyal. This is all in agreement with our model, which suggests that richly diverse cave communities evolved by the gradual addition and coadaptation of immigrant species that colonized caves at different times and places.

The almost limitless passage network of Mammoth Cave, with an array of heterogeneous cave microhabitats to suit the ecological requirements of many different sorts of troglobites, lies at the crossroads of two major subterranean dispersal corridors. One of these is the Pennyroyal plateau, which runs northward to the Ohio River and southwest nearly to the Tennessee border. The beetle *Neaphaenops tellkampfii* occurs along the Pennyroyal for most of this distance, and thus has one of the most extensive known ranges of any cave beetle.

The other dispersal corridor is a group of cave areas strung southeastward across the Cincinnati arch, forming a loose link

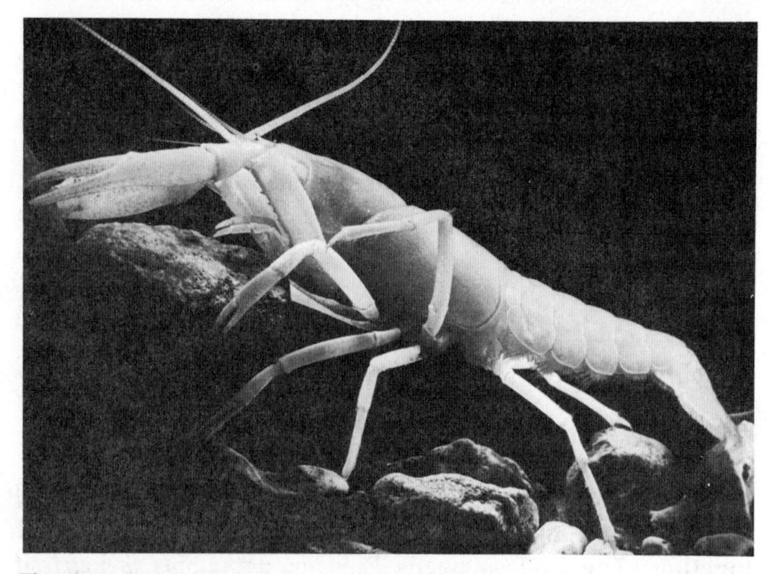

The cave crayfish, an eyeless troglobite, functions as both a scavenger and a predator. *Howard N. Sloane.*

between cave communities of the Pennyroyal and the western margin of the nearby Cumberland plateau. Nowhere in eastern North America do cave faunas approach the rich diversity of the central Kentucky karst except in the Cumberland plateau. This dispersal corridor, a sort of underground highway for troglobite immigrants, extends westward from the Cumberland plateau across south-central Kentucky and joins the Pennyroyal near Mammoth Cave.

In the Mammoth Cave system, then, we have found a conjunction of especially favorable conditions for the evolution of a troglobitic community composed of a rather large number of species. First, the cave is located near the margin of the eastern mixed forest community, some disjunct remnants of which persist in ravines and hollows in and near Mammoth Cave National Park. Conditions were therefore suitable for development of a diverse soil fauna whose members could contribute to the colonization of caves during the Pleistocene. Second, the cave system provides a vast amount of underground space, with a variety of microhabitats for species of differing niche requirements.

Finally, Mammoth Cave is at the intersection of two subterranean dispersal corridors, permitting movement of troglobites into the system from three directions—north, southwest, and southeast. Its troglobite community has gradually evolved by absorbing more and more underground immigrants, veritable living fossils whose immediate progenitors are extinct at the surface and whose very presence bears witness to the alternate waxing and waning of Pleistocene ice.

Wyandotte Cave Folk Tales

The rolling limestone hills of southern Indiana conceal hundreds of caves. The largest, and one of the most beautiful, is Wyandotte Cave. Before white men discovered it, the cave was visited and explored by aborigines and Indians. It was a source of saltpeter for gunpowder during the War of 1812. As Indiana was settled the fame of the cave increased, and by the 1850s it was described as "the world's second largest cave," exceeded only by Mammoth Cave in Kentucky.

George F. Jackson knows more about Wyandotte Cave than anyone else. And well he should; since 1923 he has been exploring, researching, and writing about Wyandotte. He himself has discovered much of the cave, and he plays a vital part in its history. He has written several books and innumerable articles about Wyandotte.

Nobody today still considers Wyandotte to be the world's second largest cave. But it certainly rivals Mammoth in its legends and stories. And George F. Jackson, alias Mr. Wyandotte, is the person to tell about it.

Wyandotte Cave Folk Tales

GEORGE F. JACKSON

I've been in hundreds of caves, but Wyandotte, in southern Indiana, is by far my favorite. I have hiked, climbed, crawled, and even dug my way through its miles of passages many times. It has been a major part of my life, and words cannot express my love for this cave. I want to share a part of this love with you and tell you some of the strange, romantic, and mysterious things that have fascinated me.

As is the case with many large and historic caves, early written accounts of Wyandotte were exaggerated. Authentic source material is scarce, and is scattered in little-known journals, letters, and newspapers. The first visitors left no written memoranda; they were aborigines and Indians who chipped calcite, flint, and other materials from the passages as long as 1,800 to 3,000 years ago. Partly burned reed torches and preserved footprints have been found in several places in the cave,

---

The Pillar of Constitution, a thirty-five-foot-high stalagmite. This forty-year-old photograph probably has been published more times than any other cave picture. George F. Jackson took the picture with flash powder, and appears in it, too. He lit the edge of a piece of paper and scrambled over to the formation before the powder went off. The Rothrocks paid young George $5.00 every time the picture was published—which may be one reason why it appeared so often. *Copyright © George F. Jackson.*

and it is from the Wyandotte Indians that the cave gets its name.

Pioneer explorers visited Wyandotte as early as 1801. In 1819 the cave and its surrounding area was purchased from the United States Government by Henry Peter Rothrock for the abundant timber on the land. As visits to the cave increased, the Rothrocks built a hotel and served the growing tourist trade. The cave remained in the Rothrock family from one generation to another until its sale in 1966 back to the government: this time to the State of Indiana. Today Wyandotte Cave is a feature attraction of Crawford-Harrison State Forest. It is managed by the Indiana Department of Natural Resources, which conducts daily guided tours.

My first visit to Wyandotte Cave was in 1923 on a Boy Scout outing. When the scouts left I stayed, spending many summers with the Rothrocks. I was a cave guide, explorer, photographer, public-relations man, and finally even a son-in-law to the Rothrocks. Since that first Boy Scout trip I have entered Wyandotte literally thousands of times.

There are hundreds of stories about Wyandotte Cave—tales of early Indian use, animal life, exploration, and many others. I wish I could tell you all of them, but of course that is impossible in this short chapter. But I can give you a sampling of some of the many fascinating legends, folk tales, and stories of the cave.

When Wyandotte's upper level of long and confusing passageways was discovered in 1858, it was only natural that it would be called "The Unexplored Regions," for it was many years before all of its passages were explored. Today the name is a misnomer as most of it has been explored, but there are some who claim that certain areas are unknown to modern man. Although it is called Langsdale Passage on modern maps, most Wyandotte devotees still refer to it by its original name. After all, it's an intriguing name for any cavernous area.

In the early days, when there was only one known connection between the Long Route and the Unexplored Regions, a trip to the "back end" from the entrance was long and tiresome. When one considers the crude lighting and homemade caving equipment used by the first explorers, it makes their trips even more remarkable. Most of the difficulties have gradually been

Tracks in the snow lead to the six-foot-high entrance of Wyandotte Cave. *George F. Jackson.*

made somewhat easier by present-day caving equipment but the early names still persist. A narrow canyon was partially filled by Rode Rock Number One and Rode Rock Number Two, long, backbone-shaped, knife-edged rocks that had to be traversed like riding on a horse. The Devil's Elbow, sticking out of the wall at the top of a narrow crevice, was extremely hard to slither past. And Wildcat Avenue, "where you have to scramble like a wildcat over a muddy floor," Andrew's Retreat, Miller's Reach, Lonigan's Pass, and some unprintable names aptly describe the hardships of the first explorations.

It was not until 1904 that Sam L. Rothrock and A. K. Sears discovered that the Double Pit in the "back end" connected with the Long Route's Air Torrent, a long, dry, sandy crawlway. This eliminated the time-consuming return trip on the upper level.

Stories about the Unexplored Regions are many and varied. Most are factual, some have an aura of mystery, and there has always been a suspicion that the "old timers"—the first explorers—penetrated parts still unknown to the rest of us. In fact, some stories have it that certain areas found by those early cavers were purposely closed by them. Why? Today no one knows.

My first visit to the "Unexplored," as the Rothrocks called it, will never be forgotten. Charley Rothrock led our party, which included some long-time guides and several Wyandotte enthusiasts who were more or less familiar with the route. It was the first trip through this part of the cave in at least ten years.

We had a grand time exploring, climbing, and crawling until we came to the Double Pit. At the top of the pit a very convenient stalagmite is located so that a looped rope may be placed around it, then pulled down after the last man has descended. We went down by sliding over the muddy edge and using a combination hand-over-hand and feet-against-the-wall method. Today, no caver would think of attempting a descent like that without a belay. Each individual got down as best he could. I was scared but, like the others, I managed to get down safely.

While we rested at the bottom someone yanked the rope down. Then Charley brought everyone up sharply by saying, "Do you realize that right here we have most of those now alive who know this part of the cave? No one has been here for years. As you know, we have to exit this area through a narrow vertical crack, the only way to the Long Route. We cannot climb back up the sheer walls of the pit. Now just suppose a slight earth movement has narrowed that crack just a few inches! We're stuck here for a long time, maybe for good!"

Although spoken jestingly, there was some truth in the statement and it immediately started a rush for the crack. Needless to say we found it just as it has been for thousands of years. I, for one, breathed a great sigh of relief.

In later years I went through the Unexplored Regions many times, often searching for the "unknown" passageways full of marvelous formations that rumor said existed back there somewhere. Eventually, I actually found one and in the finding inadvertently added another tale to the growing stories of the area.

Shortly after my first visit to Wyandotte, Charley Rothrock wrote me a brief letter, and in it he told of another trip through a section of the upper level, saying, in part, "Schontz and Yeager booh happened to get here at the same time one Saturday. They were anxious to go into the unexplored regions, so we went in early that night but only for a short distance. But we got into a region no one had ever been in before! We were in one of those high and narrow passageways and Schontz wanted to play squirrel. So he climbed up the wall for some distance, perhaps forty feet, and found an off-leading passageway. We then climbed up to him and we all traveled at least a half mile through virgin passage. We found beautiful pure white helictites and stalactites. We thought we'd photograph them on our way back but we did not return that way. Eventually we got into yet another level and finally made our way back to Langsdale Passage and the Long Route, the result being that we did not get back to the cave entrance until five A.M."

At that time schoolwork kept me from getting to the cave often, so I put the letter (which I still have) aside and, with the passing years, forgot it. Later, when I became a regular guide at Wyandotte, a chance remark reminded me of it. When I mentioned this to Bill Rothrock, Charley's son and my best friend, we decided we just *had* to find that particular passageway. The letter was at my home, many miles away, but since Charley's directions had not been very detailed anyway, we decided we could probably find the right spot without it. So, one night in we went, armed with plenty of lighting material and food. We hurried through the Long Route, crawled the tunnel past the Dead Sea, climbed the Round Room, crossed Lonigan's Pass and started looking. We climbed up narrow canyons and wide avenues but found no "off-leading passage." Just as we were about to give up we spied a small hole near the ceiling of a minor passageway. We climbed up and found the hole to be a tight squeeze, but we got in and wriggled along a dry crawlway for some distance. This was tiring work so we stopped to rest, lying flat on our stomachs. I accidentally kicked the floor. It sounded hollow. I kicked harder. Immediately a three-foot chunk of floor fell out from under me! Our backward retreat was probably the fastest ever made in the cave. Quite shaken, but curious, we carefully crawled back and looked into the hole I'd so unexpectedly opened. We saw a canyon, paralleling, but apparently not connected to, the one from which we had climbed. We estimated it to be about thirty feet deep and five feet wide. Definitely virgin, it seemed passable in both directions, but we were without ropes of any kind so we left further exploration until the "next time." As yet neither of us has made that next trip and no one else has been able to locate our accidental find. What we found that night was *not* "Charley's Passage" but an entirely new one. Some day in the future, perhaps.

My first trip into virgin cave passage was actually in Wyandotte Cave. At that time, George Jones, a Wyandotte devotee, was a frequent visitor who did a great deal of exploration in remote sections of the cave, far from the traveled routes.

One summer, for reasons he never explained, Jones became interested in the Fairy Palace Room at the end of the Long Route. At a point where ceiling and floor abruptly came together, he started a tunnel. Most people who were familiar with the cave thought this was a waste of time, but Jones persisted, usually working alone at night. This was my first year as a guide and I was fascinated with Jones' efforts to find more cave. After my guiding work for the day was over I'd trudge back to the end of the cave and help him dig. And hard digging it was: the mud was the consistency of putty, and we had only a few inches of clearance in which to work. The first man would dig out a trowel full of mud and pass it back to the man behind. He, in turn, would stuff it in any hole he could find or take it all the way out to the end of the tunnel. Night after night, as the digging progressed, our tunnel grew longer, and moving the dirt became a time-consuming problem. Eventually we reached a spot where the ceiling rose a few more inches and we managed to stuff mud into this space as we went along. Today, I have no idea how many nights we worked, but one midnight the air coming toward us seemed a bit stronger. Elated, we dug faster until we came to a place where we could almost sit up. Our lights revealed more open space through the narrow crevice ahead.

I was in front so I frantically tore the mud away with my hands. Sure enough there was a room. I squeezed through. I could actually sit upright! I wanted to go exploring immediately but George Jones' excited yells stopped me, and, reluctantly, I must admit, I helped enlarge the opening until he could slither through. We were in virgin cave, no doubt of that. Muddy, thrilled, yelling and grinning we shook hands.

What had we found? A passageway from three to four feet high and perhaps 300 feet long with a muddy floor. It ended where a block of limestone had, in ages past, dropped and filled the passage, leaving only a small space between it and the roof. A strong air current flowed over the rock and we could see open passageway beyond. But try as we might, it was impossible to push ourselves across the fallen stone; the opening was just too small. We gave up in despair and turned our attention to some breakdown along one wall. Perhaps we could get into an upper level, we thought. But there were no cracks large enough; that one room was the extent of the new find, now called "Jones' Discovery."

In later months, George Jones spent many fruitless hours trying to get beyond the fallen rock but never made any worthwhile progress. The situation remains unchanged to this day. No one knows for certain just what unknown caverns are beyond the limestone block.

When Milroy's Temple—third largest room in the cave—was discovered in 1878, it was an interesting place with some lovely formations. High on one wall, over a pit, a stream of water pours from a barrel-sized opening during the rainy season. For sixty-three years no one was able to enter this hole, then three of us managed to do it. Two stumbling blocks kept others from getting into what is now called "Jackson's Hole." Worm Alley, the crawlway from the main cave to the room, is a tight squeeze, and its sharp curves prevent any ladder-building material over three feet long from being forced through it. The other deterrent was the long, mud-covered ledge going out toward the hole. Seemingly it was held to the wall by only a tiny bit of rock. The crack between the ledge and the wall was measured in 1878 as being three inches wide. In 1941, when we made our effort, it was eight inches wide. Obviously the ledge is gradually pulling away from the wall and will someday fall.

We used some unorthodox climbing methods to reach the opening and found that another entrance, unseen from below, led down into a passage. We found Wyandotte's first waterfalls but were stopped from further exploration by a large stream issuing from an overhead crevice.

The discovery of New Cave in 1850 is undoubtedly the greatest and most important find ever made in Wyandotte Cave. We have no first-hand accounts of the original exploration, but we do have an account of how it felt to be the first caver through the Auger Hole in the fall of 1850. After the hole at the base of Monument Mountain was found, one of the men who went back to the entrance for tools was H. P. Rothrock; when he returned he also brought along his sons, Washington and Henry Andrew, who were ten and eleven years old. After hours of work the hole

Cave explorers crawl into every passage they can. Small crawlways sometimes lead to large unexplored areas. This fourteen-inch-high passage is in the New Discovery portion of Wyandotte Cave. *George F. Jackson.*

was still too small for a man to get through, but it was just the right size for the two small boys. After considerable discussion they were permitted to go in "but only for a short distance."

In a few minutes they were back, wide-eyed and scared, with tales of crawling through a damp passage to a large room with a row of stalactites across the ceiling. They could not go farther, they said, because of a tremendous hole which reached from wall to wall.

Several hours later the hole through the flowstone was enlarged so that the men could get through, and within days the complex passageways of the Long Route had been explored.

In 1921, when Bill Rothrock was ten years old, he coaxed his grandfather into taking him into the cave and showing him what happened in 1850. Bill continues with the story, "I didn't realize that my eighty-two-year-old grandfather was hardly in shape to walk all that distance and over Monument Mountain—it seemed like a short walk to me—but nevertheless he made it, even though I was irked at how slow he was. We took along a gasoline lantern and, at Grandpa's insistence, a number of candles. It took us a long while to get back to the Auger Hole. There, we rested. After a while Grandpa lit two candles and turned out the lantern. He said that was all the light they'd had when he first went through the hole, and they'd used home-made candles at that. We went on to the top of Slippery Hill, a steeply sloping hill about forty feet long that leads to a large walking-size passage; it is now dry but in 1850 it was damp and slippery; hence the name. We sat down at the top of the hill and Grandpa told me how he and Uncle Wash had crept this far on hands and knees and had stopped suddenly because in front of them was a tremendously deep pit. At first I didn't believe him but then saw what he meant. In the dim candlelight we couldn't see that there was a slope before us. It looked like the edge of a deep hole, and straight ahead there was a blackness the candlelight couldn't penetrate. No wonder Uncle Wash and Grandpa were scared and yelled for ropes and someone to come get them. I could just see those two boys sitting there, frightened almost out of their wits, thinking they were right on the brink of a black hole of unknown depth. It scared me to think of it and I realized for the first time how thrilling it must have been to have explored the cave with only crude candles for light. I never scoffed at Grandpa's stories after that experience."

Entering virgin cave isn't always easy, even when you know it's just ahead of you. An excellent example of this happened during one of my first visits to Wyandotte.

One branch of the Long Route ends at Crawfish Spring. Nearby, a passage called Wabash Avenue, which is not on the regular tour, branches off. This is the most northerly part of the cave and the area most distant from the entrance. There are few calcite formations in Wabash Avenue but there are thousands of gypsum and selenite crystals and water-carved walls, ceilings and rocks. It ends in a rather large room, Butler Point. Both the passage and the room have intrigued cavers for years but their remoteness makes them less frequently visited than other areas.

Butler Point was, however, the scene of considerable explorative activity when I visited it. Dr. Gordon L. Curry, then dean of the Louisville College of Pharmacy and a Wyandotte expert for many years, found an important lead at the very end of the room. He and an old-time guide were revisiting the area when, on the back wall, they found a small, acid-etched zinc plate that Dr. Curry had left there twenty years before. Like other conservation-minded cavers, Dr. Curry did not believe in defacing any cave by writing names or initials on walls or formations. Instead, in remote spots, such as Butler Point, he occasionally left his name and the date on tiny zinc plates that he prepared before he went into the cave.

As he reached for this particular twenty-year-old calling card he noticed a strong current of air coming from a crack behind it. Within minutes, he and the guide were digging and prying at the three-inch crack, hoping to see what might be ahead. Several hours later, a limited supply of lamp fuel forced them to stop their work.

They hurriedly made the long hike back to the cave hotel, where their story aroused great interest. That very night Charley Rothrock organized an exploring party. Suitable—and unsuitable—tools of all sorts were assembled. During my visit, Dr. Curry had "taken me under his wing" as he said, and had been explaining many scientific facts about caves and nature. Naturally, I was eager to go on the exploring trip and, after considerable persuading, Dr. Curry and Charley consented to let me go. I was even given a small hatchet to carry.

At "Curry's Crack" the group first dug a sort of pit so that the knee-high fissure could be reached more easily. They then attacked the crack with various implements in an attempt to enlarge it. And I was right in there with the rest of them, probably getting in the way, but moving rocks and dirt and tossing stones to one side as fast as I could. About every ten minutes or so someone would put an eye to the crack and try to see into it. It did enlarge a few feet back but it was impossible, even with flashlights, to see if it opened up more. The rock pounding and excavating work went on for some time. Then one of the men peered into the opening and saw a bat crawling toward him. This caused some excitement for, although bats hibernate in the cave during the winter months and some spend the summer daylight hours sleeping in "off route" channels, they are seldom seen moving through the Long Route. None of the experts in the party had seen bats in Wabash Avenue or Butler Point. The conclusion was that the bat either came from the outside or from big virgin cave. Of course everyone opted for the "big virgin cave" theory, all Wyandotte cavers believing that unexplored cave, like gold, is "where you find it."

A vigorous resumption of work was called for and when a few more bats came through, hopes rose high. Hours later, despite all the limestone that had been chipped away, we hadn't uncovered anything more than a four-inch crack in the wall. At three o'clock in the morning, Charley called it "an impossible task." I was so tired I could hardly move, but with what tools we could carry we started back.

Having brought in the hatchet, I felt duty bound to carry it back out. I finally got as far as Monument Mountain. There I dropped the hatchet and said (I'm told) "I cannot carry this any farther, I'll come back for it in the morning" and fell asleep. Charley carried the hatchet and helped support me the rest of the way out. In my room I slept for twelve hours straight and later became the butt of many good-natured jokes, for Charley had actually weighed the hatchet. When the story got around that I could not carry a one-and-a-half pound hatchet, I was sure that I'd never realize my ambition to become a guide in Wyandotte Cave!

After the digging episode, many Wyandotte specialists looked at the crack, but to this day no one has yet come up with a reasonable way to get into the passage beyond.

So the mystery remains; where did the bats come from? Surely from unknown caverns of who knows what extent.

Legends and stories about long-time commercial caves grow and grow until finally it becomes impossible to separate reality from fiction. Some early cave guides were "entertainers" as well as guides and a good exciting story added interest to a trip through the cavern. Since they were passed on by word of mouth it was only natural that some of the old stories of Wyandotte may have been embellished, for no one tells a story the same as his fellow man. A different inflection, a change of pace or tone in the speaker's voice may cause the person who later repeats the story to think he heard something he did not. So, with the following tales no one alive can possibly know just how they happened. Or did they?

Wyandotte was seldom visited before the discovery of the New Cave section in 1850. Yet it must have had a magnetic attraction for adventure-minded youths living in the vicinity. With the crude lighting materials used in those days the vast rooms and passageways must have seemed tremendous, even more awe-inspiring than today.

One day three boys from nearby farms decided to visit the cave. They rode to the entrance, tied their horses to convenient trees, and started in with only a cumbersome farm lantern for light.

The Old Cave Route, the only part then known, was extremely rough traveling. In spots it was necessary to climb over great blocks of limestone. There were deep drops where a misstep meant broken bones, or worse. In places, which have now been smoothed and dug out, it was necessary to crawl or wriggle on one's stomach. It was hard, rugged work, but exciting, fascinating, and awesome to the young explorers.

Somewhere along the route they discarded their coats. There was only one passage, so it would be easy, they thought, to pick them up on their way back. Farther along, while resting, it suddenly struck them that this would be a mighty dark place without a light. So they did the same thing that everyone visiting a cave does at one time or another: just to see how dark it really was, they turned out the light.

It was blacker than black, a darkness so intense that nothing can see in it. It sure was scary! Soon they were ready to light up. They reached for matches. With sudden shock it came to them. They had left all of their matches in their coats!

To return to their coats in the darkness was madness. There were too many deep holes they could tumble into, too many small tunnels to crawl through. Although there was only the one big passage, there were steep slopes and climbs over immense boulders, some as large as a house. They yelled, they screamed for help, all the while realizing that it was impossible for any human to hear them.

Eventually their panic subsided somewhat. They huddled together. The cave air, seemingly so warm while they were moving, began to chill them. They talked, they sang, they told jokes. Several times they started crawling. But having moved around considerably, they could not be certain which direction was the correct route to daylight.

They had heard of terrifying drops they had not passed. Suppose they went the wrong way and fell into one? That would be even worse than dying where they were.

They tried to keep track of time by counting but soon gave it up. Finally they became silent, each thinking his own thoughts of their impending doom. They were thirsty. They were racked with hunger pains. How much longer before they all passed out from starvation or some other horrible thing?

Sobbing in desperation and semiconscious, they decided that at least three days and nights had passed since they blew out the lantern. This must be the end. They could stand it no longer.

Then, when all hope seemed gone, there came a faint gleam of light! They heard voices. Rescue! It seemed impossible, yet it was true. As men came closer the boys were told that a horse had broken loose and made its way home, alerting their families.

The startled boys soon realized that their rescuers had brought no food, no litters to carry them out. In their weakened condition they couldn't walk. Surely no one expected them to make their way to the entrance without food or assistance?

Then truth came. They had been in the darkness a little more than three hours.

The power of imagination works wonders.

Two men were exploring in the far reaches of the Unexplored Regions. By way of many muddy crawlways they worked their way into an area of large passage that no one had entered before. Excitedly pressing on, they came to a branch passageway. After consultation they decided that one of them would check out the passage while the other waited. After what seemed a long time the man waiting became cold and bored and, more for something to do than in hopes of finding anything, crawled into a previously unnoticed crack on the floor. It was half filled with water but he was wet anyway and the tunnel seemed to "go" so he pushed on.

And on and on. At last he came out in a dome room. He could see no continuing passageway, except that about twenty-five feet up one wall was a ledge and above the ledge was an opening. Cavers being what they are, he climbed the slippery wall to the ledge. Sure enough, the opening looked promising.

Suddenly it came to him that he was violating all rules for sensible caving. It was time to stop and go back for his companion, pick up the rest of their gear, then return and explore the new find. But first he would have to recharge his carbide lamp. Carefully he sat down on the ledge and reached for his spare charge of carbide. As he did so he made another mistake. Fumbling around with muddy fingers he let the lamp slip out of his hand and fall over the ledge. It hit the floor below with a dull plop and went out. But no matter, he thought, he had candles and waterproof matches in his pockets. Even though it would be tough to get back down the wall with a candle in one hand he thought he could do it.

He pulled out a candle and carefully sat it upright on the ledge. But the waterproof matches? A chill went over him. He had given them to his friend when they'd separated. Now realization of his predicament came to him: alone in absolute darkness, he was in a section of the cave where no one had been before and, perhaps, in a passage whose entrance his fellow caver might never see (assuming he returned safely). He was too far above the floor to risk a jump, and climbing in the dark was almost as foolhardy.

Desperately he fumbled through his clothing for any kind of lighting material. The result was a few wet stick matches. He now had matches but no dry place or material to strike them on. Incredible as it may seem he managed to light the first match he struck.

He said, later, that he carefully rubbed a wet match head through his thick dry hair several times. When he was certain that it was dry enough he simply struck it on his teeth.

Buried gold in Wyandotte Cave? Why not? It's been buried in less likely places.

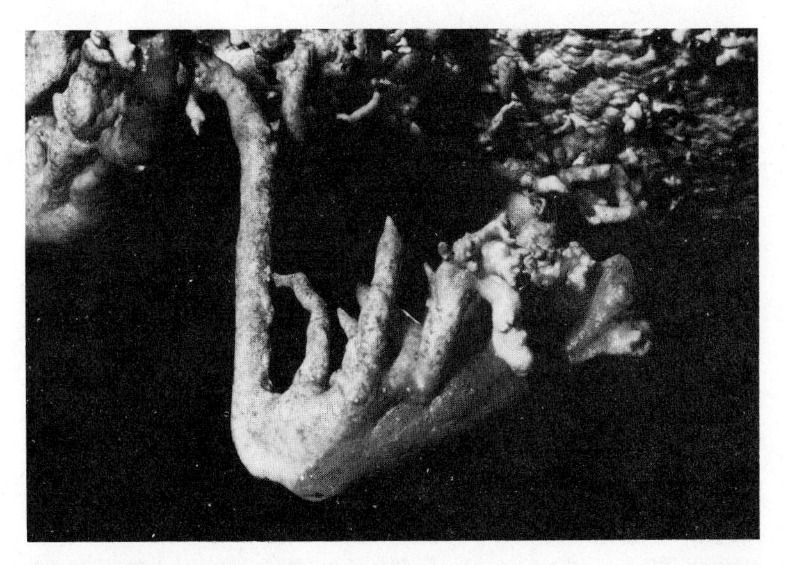

Commercial caves such as Wyandotte help protect fragile formations—like this claw-like helictite—from careless or unscrupulous hands. *George F. Jackson.*

Not far into the Short Route, just after passing through what used to be "Fat Man's Misery" (now enlarged to a comfortable walkway) and past "Bat's Lodge," a wide trench in the floor is reached. It is called "Counterfeiter's Trench." Stories about how it received that name are many and varied but the true account is possibly that given by James P. Stelle in his *The Wyandotte Cave of Crawford County, Indiana* published in 1864, the first complete book published on Wyandotte Cave. The following paragraph is taken directly from Stelle's book:

"Some years ago, two strangers came to Mr. Rothrock and after expressing themselves as being deeply interested in his cave, desired that he would give them some kind of employment in it. They appeared to be entirely destitute, having no baggage, or anything else with them; and so, out of compassion, Mr. Rothrock unlocked the entrance of the cave to them, and put

This scene in Wyandotte was taken in 1900 by Ben Haines. The man in front is Charles J. Rothrock. The legend on the original stereoscopic view was "These are pronounced the very best specimens of subterranean photography yet produced.'—Scientific American." *Ben Haines, courtesy George F. Jackson.*

them to work opening some of the narrow passages. They dug this trench and then suddenly disappeared, taking with them, as someone averred who saw them going, a large and heavy chest. It afterward came out that they were counterfeiters, and it is generally believed that their implements had, at some time, been concealed within the cave, and as it had since been locked up, they had to fall upon this plan to get them out."

Were they really counterfeiters? When I first visited the cave I asked Andrew Rothrock—who was then about seventy-five years old—and who had actually employed the men, about the story. He told me it was true and even described the men to me. Many years later I still wonder if this story is true or if the old man was just relating an interesting tale to a wide-eyed boy.

Stories of strange explorations and unusual events often occur through a mischance, a commonplace happening that is easily explained by the participants, but, when seen by others who are not aware of the entire sequence of events, it takes on an entirely different meaning. This was brought home sharply to me about 1929, when Sam L. Riely managed the cave hotel.

Guide Wally Wilkins and I went into the cave one night to check out some promising leads to a possible new cave in the South Branch of the Short Route, a section not on regular tours. As was the custom, we told Sam and others where we were going, what we intended to do, and approximately when we would return. We thought we'd be out at least by midnight.

In the early days of commercialization the South Branch was considered part of the Short Route, but because of mud here and there and because it required some stooping and retracing of steps, it was dropped and today is seldom seen except by cavers. It has some very interesting sights: Helen's Dome, high, rugged and picturesque; Diamond Avenue, sparkling with gypsum crystals; and other fancifully named spots. At its extremity a number of dirt-filled passageways branch off from the main channel; considerable excavating had been done in these spots but none of it reached open cave. Wally and I hoped to continue with digging we had started earlier and, perhaps, discover new cave.

A wealthy visitor once offered Andrew Rothrock fifty dollars for this twenty-inch-high column. Rothrock promptly refused. The column has been known ever since as "The Fifty Dollar Formation." *George F. Jackson.*

Some months earlier we two, along with Bill Rothrock, had started digging a trench. Late one night we were resting when Wally suddenly jumped up with a shout and said he had faintly heard the sound of an automobile! Although this part of the cave extends toward the main highway (at the foot of Cave Hill, one-half mile from the entrance), we never considered that we were this close to the road. For the next thirty minutes we all sat as quiet as the proverbial church mouse, listening. We heard nothing, not even a drip of water, but when we told our story it raised much speculation. Had Wally heard, possibly through a network of small tunnels, a car on the highway? No one knows, even though for many nights afterwards others followed our example and sat listening attentively and feeling foolish.

Although we talked about the auto that Wally thought he'd heard, on this particular night we did not do any listening but went to work with army intrenching shovels moving dirt from the tunnel we had started before. The mud was soft and work was slow, but we managed to push forward about ten feet before we gave up. Before we started back to the hotel we took a look at another spot of great interest to explorers: at the end of one ten-foot-high passageway further progress is blocked by a huge stalagmite that has formed completely across the passage. It is obviously the bottom of a tremendous formation that has grown here over hundreds of thousands of years and completely blocked every inch of the channel. There is only a tiny crack between the pillar and the roof. The passageway is filled as completely and as well as if tons of cement had been used to seal it. Further progress is absolutely impossible without a tremendous amount of work with heavy equipment. What lies beyond? No one knows.

By the time we had checked this frustrating place we felt it was time to go out for, after all, we had to work as guides the next day.

We were tired and muddy when we reached the hotel lobby so rather than startle the few guests still up, we went directly to Wally's car in the garage. There we curled up and immediately went to sleep. It was not our intention to sleep in the car the rest of the night, but we were more exhausted than we thought and it was daylight when we awoke. We decided to slip in the back door of the hotel, go to our rooms and make ourselves presentable for our work that day. As we walked down the hill from the garage, Sam Riely emerged from the back door and greeted us and said, "That must have been some trip if you just got out of the cave! Did you find anything?"

Our reply was a resounding "Yeah, plenty, we just got out!"

Sam looked at us with gaping mouth. It never occurred to us that we were approaching the hotel from an entirely different direction than usual.

Sam (we later learned) immediately assumed we had done the incredible and found a second entrance to Wyandotte. We brushed by him and hurried to our rooms for a quick shower and clean clothes. At breakfast we jokingly told stories of the great and beautiful rooms full of incredible formations we'd found. No one believed us. That is, no one but Sam, and it was many months before we found out that he actually believed we had discovered another entrance to the cave and had decided to keep it secret. His assumption came about because he thought we had just come out of the cave and we had approached the hotel from an entirely different direction than if we had come out the only known entrance. It never occurred to us to explain about sleeping in the car; that we had done nothing at all spectacular—that is, not until wild rumors about our trip began to float around. Our explanations to most of those familiar with the cave were reasonable and it is certain that they were believed. That is, by all except Sam. We never could convince him that we had not found a second entrance and I honestly think that until the day of his death he believed that Wally and I had uncovered the long-sought "second entrance."

One of the most persistent, yet most logical, yarns of Wyandotte Cave is that General William Henry Harrison carved his full name in what is now the Old Cave Route, when he visited it in 1806. In the early years of Rothrock ownership the signature was said to have been found, but over the years its exact location was lost. George Jones, the long-time Wyandotte enthusiast, said he located it in the early 1920s. Although somewhat interested, Jones became fascinated with other aspects of the cavern and many years passed before he was asked to "pinpoint" the name for others. By then he couldn't find it. Since the approximate location is known, perhaps it will be rediscovered in the future.

In today's well-lighted, smooth-pathed commercial caves with excellent tour guides and carefully watched visitors, it seems almost impossible that a tourist could wander off and actually get lost in the cave. Yet it has happened. Few commercial caverns admit such a possibility but most large caves have had a few lost tourists. In cases known to me it has never been the fault of the cave management or of the guides, but happened because some adventurous, foolish, visitor wanted to take a look into an intriguing tunnel or hole that the tour did not enter.

Once, while I was guiding a large party through the Short Route, one of the group, a man who had never been in a cave before, slipped off down a rough side passage not on the regular route. Armed only with a flashlight, he somehow managed to elude the second guide at the rear of the party and wandered around for some time before trying to make his way back to the rest of the crowd. He never found them, and, since the group was large, his absence was not noticed until we were back at the entrance. Only then did his friends discover he was missing. Assuring them that we would soon find him, I called Bill Rothrock and we started a search. Part of the Short Route is a circle so we thought it would be easy to pick up the man. Bill went one way, I went the other, but when we met neither of us had seen him.

Hardly believing it possible that he could have wandered into the Long Route, we nevertheless continued our search in that direction. It was at least an hour before we found him, far back in an area that was entirely different from anything he could possibly have seen, and recognized, on his tour with the group. When we approached him his flashlight had just gone out. He was covered with perspiration and could hardly speak. Tears streamed down his face and his teeth were chattering. He seemed on the verge of a complete collapse. When some composure returned we asked him if he had been frightened. "Oh, no!" he replied. "I knew where I was all the time."

Not long afterward the question came up as to how far anyone familiar with the cave could travel without light. During this particular period Charley Rothrock had a small group of Wyandotte cronies who visited him frequently, and it was decided that they would all be taken to the very end of the Long Route, searched and left, without matches or lighting material of any kind, to try to find their way back to the entrance. First, all the clothing they were to wear was checked thoroughly. We took them to Fairy Palace, watched carefully while they changed clothes and checked them all again until we were absolutely certain they had no way to make a light. Then, the rest of us dashed off toward the entrance. There was a time limit. If they were not out in fifteen hours someone was to come after them.

They made it. It took them eight hours and it was afterwards described as the toughest eight hours most of them had spent. Since Charley was more familiar with the cave than the others, most of the work fell on him. The path through part of the Long Route goes over piles of rock, and there are branching passageways here and there. Most important there was Monument Mountain. Getting off the path here would really make things tough, for in total darkness one rock feels like another, and there are holes far over to the side where one could easily fall and be seriously injured. Charley said that whenever they came to a rough spot, he felt his way along by his fingertips, trying to remember just what each rock in the path would feel like. It was hard work, but they made it and even got a story about their exploit in the newspapers. Still, that's a hard way to "see" the beauties of a cave.

Lester Howe's Fabulous Garden of Eden Cave

Lester Howe is best known for his discovery and development over one hundred years ago of the well-known Howe Caverns in east-central New York State. Howe mentioned several times another cave, the Garden of Eden, which reputedly was even bigger and more beautiful than Howe Caverns. But he never revealed the exact location of the Garden of Eden Cave, and explorers have been searching for it ever since. Some have even claimed to have found it. But Howe's description and location were cryptic. Which cave is the true Garden of Eden Cave? And, more important, does it really exist at all?

Ernst H. Kastning is a geologist with a special interest in speleology. He has published many technical articles on caves, but particularly enjoys writing about the human drama and history which so often is associated with them. He is the editor of the *New York Cave Survey Bulletin.*

Lester Howe's Fabulous Garden of Eden Cave

ERNST H. KASTNING

Howe's Cave, nestled in the rolling hills north of Schoharie County's Cobleskill Valley in east-central New York, was discovered in 1842 by Lester Howe. The ensuing development of the cave for tourists by Howe and his successors has been recounted several times, most notably by Virgil Clymer in his small book, *The Story of Howe Caverns*, first published in 1937. Today Howe Caverns is visited each year by thousands. Most of them have never heard about Lester Howe's *other* cave, the enigmatic and elusive Garden of Eden Cave.

The legend of the Garden of Eden may forever be shrouded by a curtain of hearsay, rumor, and exaggerated journalism which has persisted since the 1880s. However, several published and unpublished reports on the Garden of Eden mystery have recently come to light, and clarify several misconceptions in the oft-told story.

Lester Howe spent considerable effort and money developing and improving Howe's Cave and the surrounding

---

"Washington's Wine Cellar" as it appeared in 1889 in Howe's Cave. Visitors were shown the cave by torchlight. *S. R. Stoddard, courtesy Ernst H. Kastning.*

grounds. His expenses began to exceed the meager income derived from paying visitors, and he was eventually forced to sell the cave and relinquish control to the newly formed Howe's Cave Association. Supposedly he received a huge sum for the property, with which he purchased a farm across the Cobleskill Valley from the cave. Reputedly he became embittered by the apparent success of his former cave, watching from his farm as more and more tourists continued to visit the natural wonder. It is often told that on his deathbed he was heard to say, "I have discovered a cave more valuable than my first. I call it my Garden of Eden and have disclosed its location to no one." This purported deathbed confession sparked the northeast's longest cave hunt. Ever since Howe's death, explorers have been searching for the cave, and at times some have even claimed to have found it.

The legend underwent transformation as word was passed along from one unsuccessful searcher to the next. Additional information was often added, and some proved in error, embellishing the legend. As an example, it was reported several years ago that Arthur Van Voris of Cobleskill, New York, had rediscovered Howe's lost cave, and that he too took the secret to his grave. A victim of hearsay, Van Voris—now a former Cobleskill resident—was very much alive and happily retired in Florida. At that time I was a student and spent most weekends cave exploring in New York State. With a little work and the help of a mutual acquaintance I obtained Van Voris' address and queried him on his cave explorations during the 1920s and 1930s. He sent me several accounts of the Garden of Eden story, including both his personal writings and published information. One short newspaper article from an 1884 issue of the Cobleskill *Herald* described the farm on which lester Howe was living. He named the farm the Garden of Eden, as portions of it were sheltered from winds by surrounding hills and timber in such a manner that fifteen acres of semitropical fruit trees prospered there. The article implied that Howe was content and comfortable on his idyllic, prosperous farm. A most interesting item of note is the last paragraph:

Mr. Howe's connection with the Howe Cave project, which

> cave he discovered, did not terminate too satisfactorily for him, so he relinquished all personal connection with the cave management and ownership and went across the valley. Here it was that he purchased this acreage which we know today as "The Garden of Eden" and Mr. Howe has frequently stated that he has discovered a bigger and better cave. It is surmised by many that this is located somewhere in this "Garden of Eden," although Mr. Howe disclaims the fact.

This article appeared during Howe's lifetime, at a time when he appeared to be in good health. His claim, therefore, was not a deathbed confession.

Van Voris and his friends did most of their cave pioneering in Schoharie County during the years 1928 through 1931. In that time Van Voris recounted the group's experiences in a manuscript, "The Lesser Caverns of Schoharie County," recently published for the first time by the Mohawk-Hudson Grotto of the National Speleological Society. Many now well-known caves of the area were then first named by Van Voris's group. Several cave ventures were described by Van Voris in a small advertising booklet in 1929, *Cave Exploring with Bright Star Flashlights*, promoting flashlights sold locally by his family. Neither work sheds any light on the group's quest for Howe's Garden of Eden. However, in an unpublished 1945 manuscript, Van Voris recounted a tale of discovery by one of his comrades, Colonel Edward A. Rew:

> There was one cave, some distance away from this Garden of Eden (Howe's farm), however, which gave some promise, but which at the time of our exploration, was not accessible for more than two or three hundred feet, and during which space no sizeable rooms were found; then, at the end, the wall sloped down to a mere waterhole, with running water flowing through the hole; and we did not feel inclined to crawl into this waterway to continue the exploration.
>
> It now appears, that, according to Edward Rew, he was satisfied there was much more to this cave than we had found; so, one night, alone, he came to this cave in a dry

season; he made his way in it to the waterhole area; he then, being properly clad, crawled through the waterhole, illumination being furnished by a flashlight.

Now, inasmuch as I am writing this description and narration some weeks after it was related to me, and since I heard it only once, I cannot fill in with complete details as told to me.

However, as I recall it, Edward Rew told me that after "bellying" his way through the waterhole, he came into a more sizeable passageway leading to the north or northeast (the proper direction for our surmise). [Van Voris surely meant to say "northwest," which would be in the direction of the Garden of Eden Farm.] That he was able to continue through a continuous series of sizeable passageways, with many side channels, and with some large rooms, for a distance estimated by him to be about three miles. And he could have continued on, without too much difficulty, but being alone, with only one flashlight, he said he dared go no further, fearing possible failure of his flashlight, which would, indeed, have been tragic, to say the least.

So he retraced his steps back to the water-crawl and came out where he had entered.

This fact, as mentioned, he tells me he has told to no person.

Because of our general familiarity with size, type, direction of these caverns in Schoharie County, it appears to be more than a mere supposition that he has discovered what might well be the "second discovery" of Lester Howe, for the general direction of this cave is such that it could well have an entrance somewhere in this Garden of Eden area, toward which Edward Rew was approaching on the night in question when he made this rather exciting solitary exploration.

During 1949 and 1950 a series of three articles appeared in the Cobleskill *Times*. The first was written by Van Voris's colleague, Colonel Edward A. Rew. He had read articles in the same paper by members of the newly chartered Tri-County Grotto of the National Speleological Society, which told of their search for Howe's fabled lost cavern. Rew proceeds to needle them to get down to business, and comments that the Tri-County Grotto had already found two entrances to the cave. These undoubtedly were two small caves recently discovered in the northern part of Terrace Mountain. Rew went on to say:

> Rew's Cavern is richer in formations and is more spectacular than any hitherto unexplored cavern in Schoharie County, or, for that matter, any other in the North.
>
> I, personally and alone, on a secret expedition, discovered and explored it one night in 1931 to the extent of about two miles. I told but one person about it, a person who has kept my trust.
>
> I can say that it is in the general location of the Harold Sitzer farm, and that if they will be diligent and bold and observant of the geological formations of the countryside, they will have little trouble in locating this cavern, which they will find to be over five miles in length, complete with lake and waterfalls.
>
> They should take cognizance of the fact that The Cobleskill Valley is not as old as The Schoharie Valley.
>
> This is all the hint I will give them at this time. In wishing them good spelunking and success, my feelings are mixed, for until they do make an entrance, it's mine... all mine....

Rew mentions that he explored the cave for two miles, whereas Van Voris put his earlier estimate at three miles. It is interesting to note that Rew believes his cave exceeds five miles in length and contains a lake and waterfalls. This information becomes clearer as the story unfolds. The geologic clues in the above quotation were what later would be known as the "Finger of Geology."

In a letter to me, Van Voris related that Rew planned to commercialize the cave as "Rew Caverns" at a later date and, therefore, wanted to maintain silence on the matter. However, Rew later told his entire story to author Clay Perry in a letter, which was published in the Cobleskill *Times* in 1950:

Howe's story intrigued me, just as it has intrigued numerous others, and I started searching for this cavern, the first trip there being made by Arthur Van Voris, to this "Garden of Eden." But we found nothing of particular interest that time. I went back many times by myself, and in sitting on the rock ledges, I noted that you could see many places across the valley where Howe Caverns had had exits on different levels into the valley, and all of them pointed in the direction of "the Garden of Eden."

I also noticed that Lester Howe had built his barn at the very edge of a cliff, and had carried shale rock a good 200 yards across his land, to fill in for the barnyard. This amounted to a lot of hard labor on the part of Lester Howe, far beyond what anyone would consider necessary; so it became my belief that this barnyard either hid the entrance to his newly discovered cavern, or he wanted folks to believe that it did.

After talking with elderly local residents, I came to the conclusion that Lester Howe was an honest man and that he had undoubtedly found a cavern which was bigger and better than was his original discovery, and I realized that, alone, I could never dig away the shale that filled the barnyard.

At about this time, a gentleman living in the Town of Schoharie, came to Arthur, with the story that he had a small cave on his place, near the iron bridge on the "back road" to Schoharie village (from East Cobleskill), and he wanted us to come and look it over. We did so, and went in a short distance, when we came to a solid rock wall, with a water crawl hole at its bottom; conditions were such that although it was four or five feet wide, the water was 8-10 inches deep and the rock wall roof was only about this same distance up from the water, which would necessitate crawling through the water, holding a flashlight in your teeth.

For some reason or other, I did not feel like going on through at that time, thinking it would not amount to anything, so we returned to the entrance without going further.

Late in the fall of that same year, in the dry season, I was at home, going over my cave records and consulting the local map of that section, when I suddenly realized that all of the exits lined up as follows: on the north of the Cobleskill Valley by Howe Caverns, the Garden of Eden and this cave at the base of Terrace Mountain. So, of course, I came to the conclusion that the Cobleskill Valley was, geologically a much newer valley than the Schoharie Valley, and that, originally, the caves now known as Howe Caverns and Secret Caverns had extended on to the Schoharie River (Creek) and that the geological formation of the Cobleskill Valley had cut this originally huge cave in two, and that the larger end of the cave was left in that ridge known as Terrace Mountain, lying between the Cobleskill and the Schoharie Valleys, and that in all probability, the small section of the cave we had explored near the base of Terrace Mountain was only a passage from the now partially dead lower end of this cavern.

So I went over again to its entrance; I could easily crawl through what previously had been a "water hole crawl" (it was now dry) and I discovered a tremendous cavern. I explored it for not much over a half mile, in the general direction of The Garden of Eden, and was finally stopped by a rock cliff which had at one time been a waterfall at least thirty feet high. This cliff could not be scaled by one person, alone. It was so high that the beam of my flashlight could not reach the ceiling....

After this partial exploration, I had the choice of getting help to complete the exploration and thus letting the world know of my discovery, or, keeping quiet and preserving the memory of Lester Howe's secret, he being the only person to have ever seen this cavern.

I decided it was much more fun to keep quiet, letting Arthur, only, in on my discovery, but not disclosing even to him the exact details which I am now narrating in this article....

Now, as nearly as I can describe, the Schoharie Valley entrance location is at the base of Terrace Mountain (to the

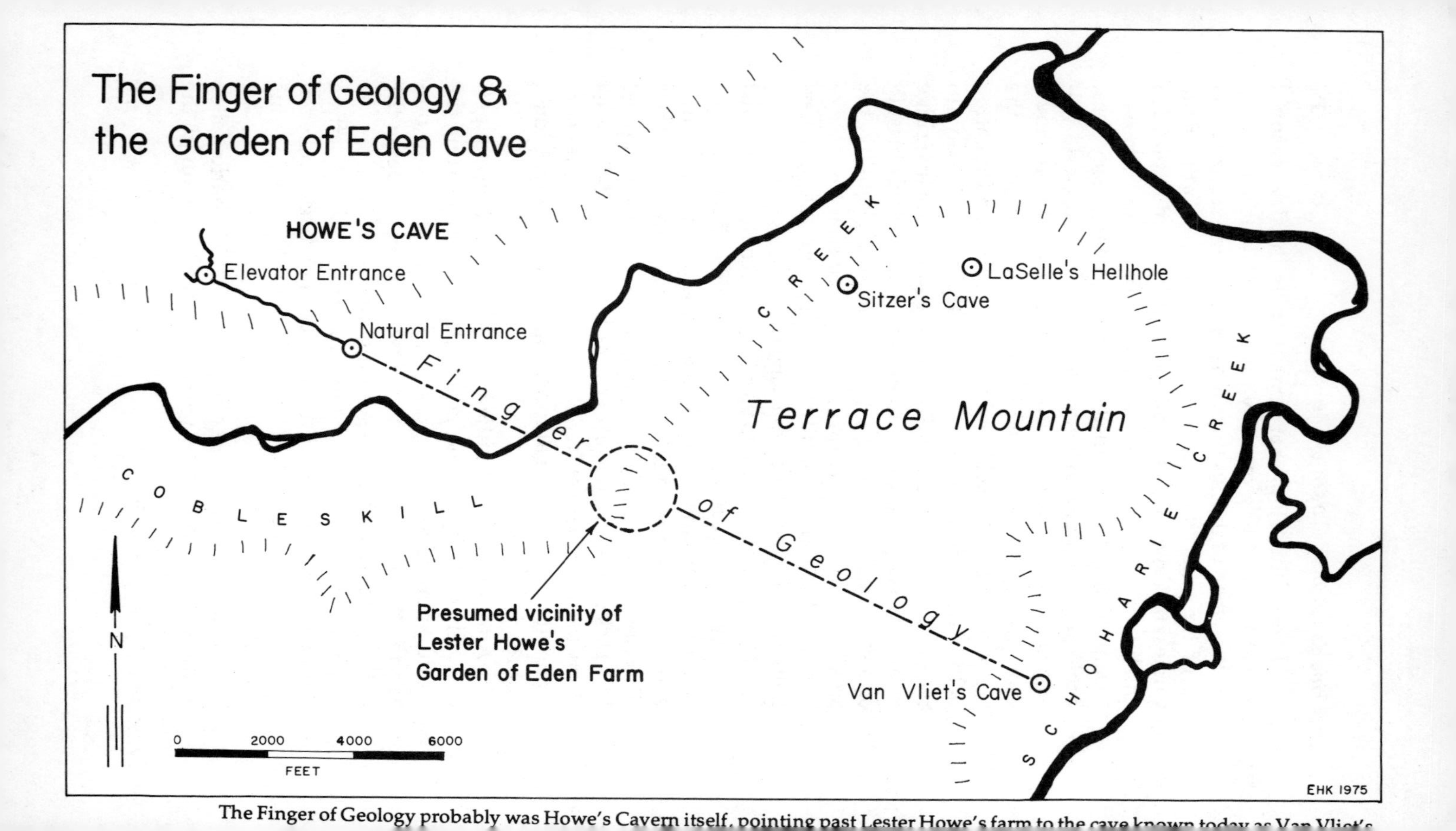

The Finger of Geology probably was Howe's Cavern itself, pointing past Lester Howe's farm to the cave known today as Van Vliet's

> cave locally known as "Van Vliet Cave"); this cave entrance is not far distant from the iron bridge over the Schoharie Creek, on the "back road," from East Cobleskill to Schoharie Village.

Rew now claimed to have explored the cave for "not much over a half mile."

In conclusion, it appears Van Voris and Rew explored a new, but small, cave on the southeast flank of Terrace Mountain, now known as Van Vliet's Cave. Rew had extended the linear trend of Howe Caverns to the southeast, across Cobleskill Valley. He postulated that the Garden of Eden Cave had to be coincident with this line on the northwest side of Terrace Mountain, and at the location of Lester Howe's farm. Rew had convinced himself that Van Vliet's Cave and two caves atop Terrace Mountain, known as LaSelle's Hellhole and Sitzer's Cave, were all genetically segments of a much larger, still undiscovered cave. The Finger of Geology, therefore, is simply Howe Caverns itself, which points across the Cobleskill Valley to Terrace Mountain, the Garden of Eden farm, and Van Vliet's Cave.

Rumors have long persisted claiming that, during 1929-1931, Van Voris and his caving group had indeed rediscovered Howe's Garden of Eden Cave. There have been at least three other groups since then who have claimed that they, too, had found the cave. Yet not one has entered a cave that meets the characteristics given by Howe.

Does the Garden of Eden Cave exist? No one knows. Perhaps old Lester Howe simply wanted to conjure up a myth which would haunt generations of cave seekers long after his death, and the entire story is Howe's joke. Even the very name itself, "The Garden of Eden Cave," suggests that, as I discovered one evening. An anagram, the words can be rearranged to spell "Vengeance for the dead."

The Ozark Underground Laboratory

Caves sometimes can be thriving businesses. Tom Aley knew that. So he bought himself a cave. He wanted to go into business for himself, and the cave business was what interested him most. But he did not put in lights, charge admission, or show tourists around.

The cave he bought was in the Missouri Ozarks—one of the greatest cave areas of the United States, renowned for its beautiful caves and unusual cave life. Aley wanted to study the cave and its life. He wanted students to come to his cave and use it as a study and research area for biology and geology. He wrote about his cave, and made educational films there. He set it up as a laboratory, showed ecologists around—but let Tom Aley, director of the Ozark Underground Laboratory, tell you about his unique enterprise.

The Ozark Underground Laboratory

TOM ALEY

The Ozark Underground Laboratory started in the spring of 1966 when I purchased the land and cave. Prior to that, however, I had spent nearly a year searching for a cave to buy, to set up my long-planned project.

The word *laboratory* may evoke a picture of a spotlessly clean room filled with test tubes, glassware, and white-smocked researchers. Such is not the case at the Ozark Underground Laboratory, where the laboratory is the natural world. This Laboratory is a 126-acre piece of the Ozarks complete with hills, valleys, underground streams, and two miles of known cave passages.

The Ozark Underground Laboratory is a unique undertaking: it has no counterpart anywhere in the world. There are other research caves, including the world-famous underground laboratory at Moulis, France. But the Ozark Underground Laboratory is more than a research cave; it is also an outdoor teaching laboratory where man's interrelationships with cave terrains can

---

Deep inside Tumbling Creek Cave Tom Aley changes the chart on the stream flow gauge station. This equipment records changes in the amount of flow and depth of the stream. *Bill Fitzgerald.*

be readily observed and discussed.

The Laboratory is not the result of a grant from a philanthropic institution, nor is it a state or federal project. Instead, the Laboratory runs entirely on money from users and visitors, and on money which I, as director, have earned working as a hydrologist. But, more important, the Laboratory exists because a substantial number of people think it should. These people have contributed labor, developed field programs around the Laboratory, and done other things to help. The Laboratory, then, is the result of work by dozens of people concerned with caves, teaching, and research.

Searching for a cave to buy was a frustrating experience. For nine months of this period my former wife Phyllis, my brother Jim, and I visited real-estate offices and looked at caves, frequently living out of the back of an old pickup truck. Some of the giant caves promised by realtors were less than 50 feet long. Others were nothing but hands-and-knees crawlways, or had been damaged by vandals, or by poor use of the land. Then there were prices inflated by legends of treasure or the mistaken belief that almost any cave could be a profitable tourist attraction if someone would just build a road to it and erect a ticket and novelty stand at the entrance. We covered much of the United States, finding that the wind blows constantly on New Mexico ridges, and that snow does fall in Wyoming in May.

In the spring of 1965 we were scouring the humid hollows of the Ozarks in Missouri and Arkansas, and I finally bought 126 acres of Missouri countryside overlying Bear Cave, near Forsyth. Since there are dozens of caves in the Ozarks named Bear Cave, I changed the name to Tumbling Creek Cave. This name was chosen because the cave stream has polished a number of chert pebbles almost as smoothly as if they had been in a rock tumbler.

After buying the cave we began the work of converting a typical piece of Ozark hill land into a center for educational field trips and research. Such work requires money, and money meant getting a job. With almost incredible good luck I found a job with the U.S. Forest Service as a hydrologist, directing studies of ground water in limestone. These findings are discussed elsewhere in this book. Money from my Forest Service salary immediately began to go into boards and cement at the Laboratory.

During the time we were searching for a suitable cave I became good friends with Delbert Walley, who lives in Forsyth, Missouri. Deb is one of those rare individuals who is truly observant of everything that goes on about him; he has amassed a wealth of information about the nature of the Ozarks. He is also a fine carpenter, stonemason, electrician, and cave-trail builder. Much of what has been built at the Laboratory, and much of what is done with the field trips, would have been impossible without Deb.

One of the early construction projects at the Laboratory was building a twenty-three-foot-deep entrance shaft. This was needed because the natural entrance, only a few feet high, required a long crawl to get to the main part of the cave. The shaft penetrates nineteen feet of solid limestone, and joins the cave in a small room at the end of a winding corridor. This location was chosen partly because it is relatively close to the surface, but also because it did not directly enter the major areas of the cave.

Caves with more than one entrance frequently have substantial air exchange. I was concerned that construction of a second entrance at Tumbling Creek Cave could drastically alter the cave microclimate, and I did not want this to occur. Since the artificial entrance is only about thirty feet higher in elevation than the natural entrance and intersects the cave at the end of a winding corridor, it did not affect the microclimate of the major parts of the cave at all.

The construction of the shaft did slightly alter the temperature and relative humidity conditions of about 200 feet of cave passage near the entrance. The change is noticeable only in the summer, when surface temperatures are higher than the cave's average temperature of 58 degrees. When the surface temperature is warmer than the cave, cold air sinks out of the cave (since this air is denser and thus heavier) through the natural entrance, which is lower in elevation than the shaft entrance. As the cold air moves out, replacement air is drawn into the cave through the shaft entrance. This air is warmer than cave temperatures, and thus warms the cave slightly. Much of the time the dew-point temperature of this replacement air is higher than 58 degrees, so that moisture condenses on the cave walls near the shaft. At times, however, the replacement air has dew-point temperatures which are less than 58 degrees. When this air enters the cave, it picks up water from the cave. Almost all of this water exchange occurs in the first two hundred feet of the cave, and on balance tends to dry out this portion. To keep the air exchange to a minimum, we put a building over the shaft and built an air-flow reduction door in the cave about 25 feet in from the bottom of the shaft.

The direction of air flow in the cave is different in winter than in summer. In winter, the cave air is warmer than surface air. Warm air rises from the upper entrance of the cave (the shaft entrance), and replacement air is sucked in the lower, or natural, entrance. Before we built the door in the cave and finished the building over the shaft, we went to the cave one bitterly cold Saturday morning. The surface temperature was about two or three degrees Fahrenheit, and great quantities of air were gushing from the shaft because of the great difference between surface and cave temperatures. As the air left the cave it cooled rapidly. Warm air can hold much more moisture than cold air. In this case, the cave air could hold about 12 grams per cubic meter more moisture than could the surface air. When the cave air poured from the shaft and was rapidly cooled to surface temperatures, moisture condensed and formed delicate crystal festoons of hoarfrost on every tree limb and bush around the shaft, and on branches as high as thirty feet in the air above the shaft. It was a beautiful but frigid scene as we climbed down into the cave and into the moist warmth of the underground.

At the Laboratory we seek to maintain the cave in as nearly a natural condition as possible. A simple philosophy, perhaps, yet for it to succeed, the intrusions of man must be minimal. But intrusions for education, knowledge, and appreciation are the purposes of the Laboratory. We thus face the classic problem of protecting and preserving wild places, but still making them available to people. The problem is how to intrude gently.

The surface is far more capable of recovery from intrusion than is the cave. On the surface, the rain and wind erase footprints; in the cave, footprints in the mud can last for years, or for hundreds of years. In Tumbling Creek Cave I have carefully threaded my way through stalactite-studded crawlways only to find tracks and clawmarks of bears in the passages beyond. These marks date from times when there were other natural entrances to the cave system, and from times before the stalactites grew and almost barred the passage. The footprints I make, unless obscured by footprints of subsequent people, are destined to far outlive me.

Surprisingly, one of the most damaging things people can do to caves is to walk through them. The typical Ozark cave has footprints everywhere. Tracks lead across the flowstone, where they discolor the crystals. Footprints traverse moist mud floors, where they change the environment to a swamp resembling a surface cattle-watering area. And the tracks go from tilting rock to tilting rock as they progress along the underground streams. Beneath some of those tilting rocks, should you turn them and look, are Ozark blind salamanders, *Typhlotriton spelaeus*. These salamanders found benath the rocks are on the national Rare and Endangered Species List. I wonder how many are unwittingly squashed by lunging cavers.

To avoid the problems of trampling at the Laboratory, we began building trails in the cave as soon as we finished the shaft. All traffic is kept on the trails, narrow paths made of cement mixed with the cave mud. We carried in the cement in packs; there is nothing quite like backpacking through a cave with a 50-pound load of cement. Everything off the trail is as natural as we can keep it. The trail construction took Deb and me, and sometimes Deb's son Bill and my brother Jim, five years to complete. We worked almost every day of every weekend.

We have built about 2,100 feet of trail through Tumbling Creek Cave. The trail passes through sections which are representative of almost all conditions and environments found in the cave. The average visitor sees about twenty percent of the cave. The remaining eighty percent has no trail, and we anticipate building no more pathways.

There are times when we travel beyond the limits of the trail. Such trips are made with small groups and for very specific purposes. Even in the undeveloped portions of the cave, we restrict walking to a narrow zone. Where we have discovered new sections of the cave, all our travel has been in narrow zones, and much of the floor is totally untracked.

Animal life in Tumbling Creek Cave is, as caves go, very abundant. At present we have at least preliminary identifications of about seventy species of animals. These include bats, salamanders, frogs, crayfish, spiders, several insects, isopods, and others. Many of the animals found in the cave are troglobites, or true cave dwellers, those that cannot survive in surface conditions.

Animal life in caves is usually very limited, in large part because of a great scarcity of food. Some food is washed into Tumbling Creek Cave, but the main source of food is bat guano.

During the summer a maternal colony of gray bats, *Myotis grisescens*, inhabits the cave. During the period from May through October there are about 150,000 bats using the cave. Many young bats are born during this time. The bats leave the cave at night as soon as it reaches full dark outside. It is an unique experience to sit at the entrance and watch 150,000 bats fly almost silently from the blackness of the cave into the darkness of the night. On moonlit nights it is relatively easy to see the bats, but on overcast or rainy nights they are nearly invisible, and one more properly senses their passage overhead than sees it.

Upon leaving the cave, most of the bats fly toward Big Creek, where they apparently drink before starting their bug-catching in earnest. Each bat eats approximately half his body weight in flying insects every night; a gray bat weighs between 1/3 and 1/5 of an ounce. The bat colony thus eats about 1,000 pounds of bugs a night. A thousand pounds of insects a night; yet many people who have never seen a bat consider them repulsive creatures. Bats must rank as one of the most under-appreciated animals on earth.

After feeding, the bats return to the cave. The indigestible residue of their meal is excreted and deposited in the cave as guano, a black manure rich in nitrogen. The guano is scattered wherever the bats fly, but much of it is found beneath roosting sites. The gray bats typically hang in dense clusters on the cave ceiling; densities are frequently 150 to 200 bats per square foot of cave ceiling. That means a lot of guano, and a lot of energy input into the cave system.

In 1973 Chris Bashor, a high-school student from Fayetteville, Arkansas, conducted a science fair project at the Laboratory. Using a bomb calorimeter, he determined the caloric value of bat guano from Tumbling Creek Cave. He found that bat guano in the Big Room has about 3.5 calories per gram. That is roughly the same caloric value as sirloin steak.

It is the bat, then, that provides the bulk of the energy input to the cave food chain in Tumbling Creek Cave. The bat is the catering service, bringing in food and distributing it as he flies through passages of the cave. Without the bat and his vital catering activities, other animal species in the cave would be much less abundant, and some species might totally disappear. It is thus imperative that we understand and appreciate the important function of the bat in cave ecosystems (and for that matter, in surface ecosystems as well; 1,000 pounds of bugs a night is nothing to scoff at). The bats need protection; the gray bat, the very species that is so vital to Tumbling Creek Cave, is on the Missouri list of endangered wildlife species.

Why should a bat species with 150,000 creatures in Tumbling Creek Cave be on an endangered species list? The problem is not primarily one of a limited number of bats, but rather of a limited and ever shrinking amount of habitat. It is a rare cave that provides the conditions necessary for either maternal or hibernating colonies of gray bats. There are probably not more than five or ten important maternal sites in Missouri, and probably only half that number of hibernation sites.

Furthermore, if gray bat colonies are disturbed by people, the bats may quit using the cave. Disturbance of maternal colonies in the late spring, or disturbance of hibernating colonies in the winter, may cause substantial bat mortality. As more people visit the cave, disturbance is an ever-increasing threat to the gray bats. Several caves important to the bats have been developed into tourist attractions, and this has led to substantial expatriation or even extermination of these bats. In one of the caves the management burned a stack of old boards in the cave. Thousands of bats perished in the smoke.

Among the seventy species of animals found in Tumbling Creek Cave are eight species new to science. Among these new animals is the Tumbling Creek snail, which is not only a new species, but a new genus as well. The scientific name for this tiny snail is *Antrobia culveri*. Its entire known range consists of a few hundred feet of the underground stream in Tumbling Creek Cave. This, the total known range of this animal, is smaller than the average suburban house lot in the United States. This limited range led to inclusion of this snail on Missouri's list of endangered wildlife species.

The Tumbling Creek snail has a beautiful white shell about the size of a grain of sand. They are very particular about where

The Tumbling Creek snail, *Antrobia culveri*, known only from a small portion of the Laboratory cave and no place else. This tiny gastropod is about 2 millimeters in diameter—less than one-tenth of an inch. *Bill Fitzgerald.*

they will live. The snails can be found attached to the bottom surfaces of rough-textured rocks in the cave stream. Snails are most easily found on slabs of chert coated black with the mineral hausmannite, a manganese oxide. Snails are not found in those portions of the stream where the bottom is mud or small gravel, and they are not found upstream of the areas used by the bats.

I feel very protective toward my snails. They are unique pieces of life, just as much so as the wide-mouthed rhinoceros of Africa or the elephant tortoise of the Galapagos Islands, which are also endangered species. If one is concerned for the protection of the rhinoceros and the tortoise on ethical or scientific grounds, I think one should also be concerned for a tiny white snail.

At first it might seem an easy task to protect the snail. I do not collect them, nor allow others to do so. I lock the cave when I leave. These safeguards may be important, yet the fate of the snail will ultimately be decided beyond the boundaries of his range and out of my control. Should the bats lose their fight for survival, be expatriated from the cave, or even be greatly reduced in number, the snails would lose their major source of food and might well perish. Or what if a septic-tank leak or misapplied dose of pesticide found its way to the cave stream from a surface sinkhole or stream? The Tumbling Creek snail has a very tenuous hold indeed on his tiny piece of this earth.

Many of the activities at the Ozark Underground Laboratory are tied to our educational field programs. Basically we conduct three types of programs. First, most of our efforts are directed toward one- and two-day field sessions for college and high-school groups, and occasionally for junior-high groups. Most of these sessions are on weekends, and many of the groups visiting the Laboratory are from 200 to 300 miles away.

The second type of program runs for several days at a time. These sessions typically involve student-conducted investigations. We try to develop in the student deeper insights into ecosystem functioning than is possible with the one- and two-day field trips.

The third type of educational program is directed toward

groups which may never have heard of the Laboratory and probably will never visit it. The narrative you are reading is one example. We have also made several films, both for schools and for the general public, at the Laboratory.

Much of the emphasis in our one- and two-day field trips is on relationships between the surface and the cave. What happens on the surface affects the cave, and, conversely, what happens in the subsurface affects the surface.

To understand the land in a cave region, one must not think of the terrain as a two-dimensional surface. Instead, to truly understand the landscape and what occurs on it (and in it), you must visualize both the surface and the subsurface as well.

Instructing student field trips is an important function of the Ozark Underground Laboratory. Tom Aley explains a stalactite drippage station to a class. Water dripping from a stalactite runs through tubing into the barrel; the recorder on top of the barrel measures the rate of water accumulation. *Bill Fitzgerald.*

As we stand on the earth, *up* consists of air, and *down* consists of soil and rock. On the earth, length and width are far more similar than up and down. Length and width each have a separate dimension, but up and down are expected to share the vertical dimension of air (up) and earth (down). When we deal with the earth I believe we should call *up*, the air, the third dimension. As for *depth*, I like to think of it as the fourth dimension. At the Laboratory it is this fourth dimension of soil and rock which we emphasize, but we tie it to the other three.

Most field trips at the Laboratory consist of two segments: the surface tour and the underground tour. On the surface trip we discuss some of the ways in which the surface affects the cave. For example, much of the water entering the cave system enters through very localized surface areas. Sinkholes and losing streams (surface streams which lose water into the subsurface) are two important classes of localized points of groundwater entry. On the surface trips we discuss these areas as potential sources for groundwater contamination. Through studies we have conducted we can show that water from sinkholes and losing streams enters the cave stream within a matter of hours, and that it does not receive effective filtration or other natural cleansing as it oozes through soil and rock.

The cave portion of the field trips emphasizes relationships of the cave to the surface. Many people view caves as a world which is unique and somehow distinctly separate from anything on the surface. But the cave systems and the land mass in which the caves lie are intimately connected; they are not separate worlds. Failure to consider the effects of surface activities on underlying caves can lead to distinctly unfortunate consequences. For example, asphalt parking lots for commercial caves have been paved directly above the very cave that is the attraction. These parking lots have seriously interfered with water drippage of stalactites in the underlying cave system. This has damaged the formations and made them less attractive for visitors. The surface affects the subsurface.

At the Laboratory, we try to relate the features we see in the cave to events on the surface, or to processes which are occurring

Deb Walley (left) and Tom Aley examine a small stream in Bear Cave Hollow. This stream disappears underground in front of the flat rock at the bottom of the picture to join the cave stream in Tumbling Creek Cave. *Bill Fitzgerald.*

between the land surface and the cave. For example, there are some massive old stalactites and stalagmites in the cave which no longer receive water. Both in size and form they differ from the speleothems which are forming elsewhere in the same chambers today. The massive formations are related to past conditions in the area, and distinctly not to the conditions of today. Although our investigations have not progressed to the point where we can characterize the conditions which existed in the area when these formations were building, we are able to show dramatically that changes have occurred, and that conditions in the cave have changed with time.

The food chain within the cave system provides us with an excellent way of relating the cave to the surface. As was mentioned before, the primary source of food for the cave comes from bat guano derived from outside insects. Were it not for the outside insects being eaten by the bats, there would be far less available food for the cave animals.

In Tumbling Creek Cave the Ozark blind salamander is typically regarded as the top animal in the food chain. The salamander, which as an adult is about three or four inches long,

Wood washed into the cave during heavy stream flow supports a half-inch high mushroom. The wood and fungus are an important part of the food chain, second only to the bats, in providing nourishment for many cave animals. *Bill Fitzgerald.*

is preyed upon by no other cave-dwelling animal. It eats insects and other invertebrates drawn by the bat guano. However, during the winter of 1973 we noticed a very interesting occurrence. A raccoon entered the cave, most likely through the natural stream entrance, and travelled through at least 1,000 feet of the East Passage. The raccoon was thus 2,000 to 3,000 feet from the entrance, which in complete darkness is a rather impressive feat in itself. But more impressive yet is the fact that the raccoon walked through essentially every pool in the East Passage. Salamanders are typically found in these pools. Before the raccoon's visit we would always see one and usually several salamanders in the pools in the East Passage. For about three months after the raccoon's visit we saw salamanders only on about half of our trips into this part of the cave. Even when we saw salamanders, we usually found only one. From this I conclude that the raccoon ate many of the salamanders in the East Passage.

An immature Ozark blind salamander, *Typhlotriton spelaeus,* has external gills and functional eyes. This one, which is nearing metamorphosis to become an adult, will soon lose both gills and eyesight. *Howard N. Sloane.*

If it did, we have a number of very interesting questions. How did it navigate in the total darkness of the cave? The raccoon made such an apparent impact on the salamander population that I am convinced it is a very effective salamander hunter. The capture of salamanders was not "pure luck"; the captures depended on skill and effective hunting techniques. An intriguing question is how the raccoon stalks and captures salamanders. What senses does it use to discover and capture the salamanders?

The questions about the raccoon could go on and on, but the important point for us is that it is a surface animal which is apparently at the very apex of the cave food chain, preying off the Ozark blind salamander. The base of the food chain, bat guano, is also derived from the outside. Who could look at food relationships such as this without concluding that the surface and the cave are intimately connected?

Field school programs at the Laboratory usually last from three to seven days. In these programs everyone conducts his own investigations, and, hopefully, develops a more personal comprehension of the ecosystem at the Laboratory. During the summer of 1974 we began week-long field schools through the Memphis Museum for junior-high and high-school students from Memphis, Tennessee. Marcia Kelly, an ornithologist with a good botanical background, joined the Laboratory staff to help develop field-school programs. If the reaction of the participants is any indication, field schools will undoubtedly be an important Laboratory activity in the coming years.

The third type of educational program at the Laboratory is one of educating the public at large. This involves lecturing, writing, and working with educational films and television. We want to help people understand how ecosystems in cave regions function, and how man must relate to these systems to keep from damaging or destroying them.

There is an axiom in sociology which maintains that what is perceived is real, for it has real consequences. I like to apply the same axiom to land use in cave terrains, although it of course applies to any terrain.

Decisions on land use in cave terrains are based upon the understanding (or perception) a person has of how the system functions. If this understanding is incorrect, then the decision may also be incorrect. Incorrect decisions in land use may result in undesirable consequences. Let me give an example.

If people think the subsurface in a cave terrain naturally purifies water, then they will put wastes underground with little concern about where they are placed. Sinkholes will be used for dumps, and septic systems will be located near wells and in the recharge areas of springs. This is exactly what has happened in the Ozarks. Unfortunately, this idea is completely wrong. The subsurface in the Ozarks does not provide effective natural cleansing for water. Rather than the subsurface functioning as a nearly infinite cleansing mechanism, it more correctly functions as an elaborate plumbing system which transfers problems from one area to another. Thus sinkhole dumps and septic tanks have contaminated springs and wells throughout the Ozarks.

Salamander tracks and tail marks in a mud-buttomed pool at the Laboratory. *Bill Fitzgerald.*

Because of our concern with protection of groundwater in the cave regions of the Ozarks, the Laboratory became deeply involved in a 30-minute television documentary, "Please Don't Drink the Water," produced by Station KYTV in Springfield, Missouri. Most of the program focused on activities of the Laboratory staff and on work conducted at the Laboratory. The program was shown during prime time on a Sunday night, and has since been shown on several other stations and to numerous groups throughout the country. I hope the program helped people understand how serious the hazards of groundwater contamination are in cave terrains.

The Laboratory was also involved in the filming of the National Geographic Society television special, "Strange Creatures of the Night." The film crew came with 50 boxes and cases of photographic equipment which must have weighed about 3,000 pounds. It had to be carried, pushed, and cursed for about 1,000 feet into the cave, and then back out again.

The performers for this portion of the program were Dr. Bob Mitchell from Texas Technical University, Dr. Tom Poulson from the University of Illinois, and Deb Walley and myself from the Laboratory. No studio chairs with stars on the back for us; instead we served as equipment haulers. And yet perhaps the most taxing part of the entire filming for us was the day spent in my Land Cruiser getting surface road scenes.

A Land Cruiser is not a particularly large vehicle, yet we packed in our four-man scientific contingent, plus a cameraman, a sound man, and a director-producer. And then, so that our conversations could be heard on the sound track, we rolled up the windows. My car does not have air conditioning, yet we spent much of the day driving through the southern Missouri August heat in our hermetically sealed sardine can. In deference to our perspiring faces, the director-producer would periodically yell "Wipe!" and we would all take out paper towels and dry our faces. But the final blow came when the program was shown. There were no road scenes. They had all been cut.

In addition to the television documentaries, the Laboratory has also been involved in making educational films. The film "Cave Ecology," produced by Centron Corporation and directed by Dr. James Koevenig, was filmed at the Laboratory. Designed for elementary school use, it shows a number of the animals found in the cave.

The Laboratory has also been featured in a number of articles and books. As an example, a book on the Ozarks in Time-Life's American Wilderness Series devotes twelve pages to the Laboratory. But perhaps more important than the publications are the lectures given by the Laboratory staff in distant schools. We give thirty to forty lectures a year to high-school assemblies and college classes in the Midwest. I get hoarse just thinking about it.

Over sixty percent of Missouri is underlain by limestone and dolomite, soluble rocks in which caves form. Similar soluble rock terrains constitute a substantial part of the United States, and it is essential that we learn to live harmoniously with these lands. If people are to learn this, then it will requre study and teaching. And that, stripped to its bare essentials, is what the Ozark Underground Laboratory is about.

Little Alice and Lost John

The human history of Kentucky's Mammoth Cave goes back to the Indians and aborigines who explored the caves of the area with reed torches. Radiocarbon dating of relics indicates that human visitors trod Mammoth's passages more than 3,000 years ago. A few of these early explorers met with accidents underground, and the bodies of others may have been placed inside caves after death. Some of these were preserved and mummified by the cave atmosphere.

When the Mammoth Cave area was settled adventurers began to explore and map the many caves. Tourism increased markedly after the War of 1812. Two mummies were found in Mammoth Cave, and others were discovered in nearby caves. Since there were many caves open to the public, the cave operators were quick to utilize all forms of advertising to entice visitors to their cave. Some of the main attractions became the mummies.

Harold Meloy first visited Mammoth Cave in 1925, and he has been fascinated by the cave's history, legend, and folklore ever since. His extensive writings about Mammoth Cave history include a book about the mummies, and he has helped the National Park Service to interpret accurately the history of the cave to the public. He is a director of both the American Spelean History Association and the Mammoth Cave National Park Association, and is a Fellow of the National Speleological Society.

Little Alice and Lost John

HAROLD MELOY

A mummy has been a part of the Mammoth Cave story since tourists first visited the cave during the War of 1812. Seven mummies have been found in the caves of the region. Here are the stories of two of them; "Little Alice," a mummy discovered in nearby Salts Cave in 1875, played one of the stellar roles among the small mummies of Mammoth Cave, and "Lost John," an Indian miner, can be seen by today's cave visitors, protected by a glass case.

By 1813 an Indian mummy had been discovered by saltpeter miners in a cave a few miles south of Mammoth. The following year two more mummies were found near the crypt of the first. Appropriately, the cave became known as the Mummy Cave. Also known as Short Cave, it was all but forgotten within a few years after the mummies were taken away.

The first mummy was taken to Mammoth Cave, where it remained until the autumn of 1815. The second mummy was sent to New York City and exhibited by John Scudder in his Amer-

---

Audubon Avenue, discovered by Mammoth Cave saltpeter miners in the early 1800s, was used earlier by Indians to reach the gypsum deposits. *Photo by W. Ray Scott. Courtesy National Park Concessions, Inc.*

ican Museum. The third mummy may have been brought to Cincinnati.

Hyman Gratz, a Philadelphia merchant who was co-owner of Mammoth Cave, was a friend of Charles Willson Peale, the great American portrait painter and the proprietor of Peale's Museum. There is every indication that Gratz intended the mummy in Mammoth Cave to find its final resting place in Peale's Museum in Philadelphia, but this was not to be.

Nathan Ward, a native of Shrewsbury, Massachusetts, had other plans for the mummy. By 1815 he had probably learned of the "Mammoth Cave Mummy" on exhibit at Scudder's museum and had undoubtedly read in the scientific press of the day that there was still a mummy at Mammoth Cave intended for Peale's Museum. Ward journeyed to the cave via Lexington, Kentucky, where he talked to Charles Wilkins, the other co-owner of Mammoth Cave. When Ward arrived at the cave in October 1815, he told manager Archibald Miller that he had Wilkins's permission to take the mummy. With a good eye to business, Ward made an eighteen-hour inspection trip in the cave, conducted by two competent guides, and on his departure took the mummy with him.

Then Ward wrote and published a long descriptive account of his explorations in the cave and his discovery of the mummy in one of the passages. The public was curious to see the mummy (for a fee), and business was good in the cities and towns in which he sojourned en route back to Massachusetts. Ward's published account describing the cave had wide reader appeal. It was published and republished in whole or in part in numerous newspapers, magazines, and books in this country, and later appeared in England and elsewhere in Europe.

Some of these publications were illustrated with a picture of the mummy, and the "Mummy of Mammoth Cave" became

---

Scudder's Mummy was discovered 1814 in Short Cave, and exhibited from 1815 to 1841 at Scudder's American Museum in New York City. The museum, purchased by P. T. Barnum in 1841, burned in 1865. Scudder's Mummy is frequently called the Mammoth Cave mummy. Drawing made circa 1816 by Constantin S. Rafinesque and published 1817 in the *Medical Repository*. *Courtesy of The Historical Society of Pennsylvania, Philadelphia.*

famous in current popular literature on both sides of the Atlantic. By 1817 the American Antiquarian Society of Worcester, Massachusetts, had obtained the mummy from Ward. By 1820 there were three different Mammoth Cave mummies on display to the curious in three different museums: Worcester, New York City, and Cincinnati. It was primarily Ward's published descriptions of the cave, sometimes with a map of the cave, reprinted numerous times in numerous cities—and the mummies—that made Mammoth Cave famous.

More and more travelers came to the cave. But there was no mummy to show them. Soon the guides began directing the attention of the paying customers to the place where the mummy had been, and they explained that in the early days it had been discovered there "in a state of perfect preservation." For half a century the "Mummy Seat" was one of the points of interest to be seen on the regular cave tours. And the story improved with the telling. Now the mummy was an Indian princess or a queen buried with all her regal clothing, deer skins, beads, necklaces (one made of deer hoofs), jewelry, and other accessories.

Among the many visitors fascinated by this story was Nathaniel Parker Willis, who christened the missing mummy "Fawn Hoof, Kentucky's posthumous belle." The story continued to grow with the retelling by successive guides. According to another version, when the mummy was discovered it was accompanied by a mummy of a child, sitting in its own smaller mummy seat near the niche which accommodated the adult mummy. Dr. Charles W. Wright, in recounting the story, wrote that "they were both in a state of perfect preservation. There can be no doubt but they wandered into this avenue, and becoming bewildered, sat down and died in the position in which they were found."

Fawn Hoof, an Indian mummy discovered 1813 in Short Cave, was often designated as the original Mammoth Cave mummy. It was exhibited in Mammoth Cave, 1813-1815; at the American Antiquarian Society, Worcester, Massachusetts, 1817-1876; and then at the Smithsonian Institution until circa 1900. Photograph made at Washington, D.C., prior to 1896 and forwarded by Dr. G. Brown Goode to Dr. Richard Ellsworth Call. *From Hovey and Call: "Mammoth Cave of Kentucky," 1912.*

But in 1870, when Dr. William Stump Forwood published his authoritative book, *Mammoth Cave of Kentucky*, he indicated there were "doubts regarding the whole question." This was not good for the lucrative tourist trade, and in the 1870s, Mammoth Cave needed a mummy badly, not just a legend, but a real mummy to perpetuate the story of Fawn Hoof and to excite the interest of the underground customers. Such a mummy was soon found in Salts Cave, but it was spirited away before it could fill the immediate need at Mammoth.

Salts Cave was one of several caves in the vicinity owned by the Mammoth Cave proprietors. Tradition held that it was almost as large as Mammoth and that there probably was an underground connection between the two. With or without permission of the owners, some of the local people explored the seemingly never-ending passages of Salts. It was fascinating; they never knew what might be beyond the illumination of their feeble lights; and it was possible that they might find the connection with Mammoth—or even a better cave than Mammoth. In either event the financial rewards would be promising, especially if a new discovery happened to lie under land not owned by the Mammoth Cave proprietors.

They did find numerous Indian artifacts similar to those found in Mammoth, and in 1875 two of these explorers found another Indian mummy. But they knew that there would be no glory and little if any financial reward for their discovery. For the Mammoth Cave owners would immediately claim the property belonged to them; they needed such a mummy to play the role of Fawn Hoof, the "original Mammoth Cave Mummy."

But there was a more promising alternative. Larkin J. Proctor had formerly been the manager at Mammoth Cave, but had recently lost his lease. And Proctor owned some caves of his own in the neighborhood, including Proctor's Cave, which he was showing commercially at the time. The newly found mummy was secretly removed from Salts Cave and deposited in one of the caves owned or leased by Proctor—by local tradition in the cave known first as Wright's Cave, later as Long Cave. Then, when the mummy was "officially" discovered, it could be the property of Proctor; and the exhibition fees need not be accounted for to absentee owners of the Mammoth Cave lands.

This mummy was smaller than Fawn Hoof, and like the others, was a desiccated body naturally mummified by the dryness of the cave environment. Later called "Little Alice," she was an immediate success. Proctor exhibited her as the cave mummy at his cave, and later she was shown at the Long Cave, renamed for commercial reasons Grand Avenue Cave.

Henry C. Ganter became the next owner of Little Alice. In the 1880s Ganter had managed Mammoth Cave; during the 1890s he managed both the Mammoth Cave Hotel and the cave; and he continued to manage the cave well into the twentieth century. He brought Little Alice to Mammoth Cave and exhibited her there as the Mammoth Cave mummy. This pleased the customers, and Ganter saw no reason to explain that this was not the original mummy found many years before. Rather, he encouraged the efforts to bestow it with all of the history, folklore, and legends formerly associated with Fawn Hoof.

There was little competition. The original Fawn Hoof was acquired by the Smithsonian Institution in 1876, which exhibited the mummy at the Centennial Exhibition in Philadelphia. Subsequently it was shown in the old National Museum in Washington, D.C.; but the plaque stated that she had been discovered in Short Cave. The museums housing the other two mummies in New York and Cincinnati had each been destroyed by fire decades before. At least two scholars must have known that the present mummy (Little Alice) on display at Mammoth Cave was too small to be Fawn Hoof. Dr. Richard Ellsworth Call, a careful investigator and one of the more interesting writers about the cave, obtained a photograph of Fawn Hoof from Washington, but he did not publish it with his story about the mummy. Probably Ganter persuaded him not to, for a published picture of Fawn Hoof which could be held beside Little Alice would clearly show that they were two different mummies. Reverend Horace C. Hovey, a recognized authority since 1882 on all matters pertaining to the cave and a confidant of Ganter, must also have known of the substitution, but his natural chivalry where ladies were concerned may have restrained him from exposing Little Alice's secret—and Ganter's.

Little Alice, an Indian mummy, was discovered 1875 in Salts Cave and shown intermittently at Mammoth Cave from about 1890 to 1960. *Drawings by Marilyn DeLaurentis.*

Little Alice had great publicity value. Ganter sent her to exhibitions to promote business for the cave. Local oral tradition holds that in 1893 Little Alice was sent to the World's Columbian Exposition in Chicago, together with some of the beautiful gypsum flowers from one of the cave passages, and accompanied by cave guide William Bransford.

However, in 1897, Call wrote that the mummy (Fawn Hoof) which had been at Worcester was exhibited at the Philadelphia Centennial Exhibition and then became the property of the National Museum in Washington. It was to Ganter's interest to merge into one the identities of the two mummies, and he let it be known that he, too, had exhibited his mummy (Little Alice) at the Smithsonian Institution. Fifteen years later, Call's text was changed to read that the mummy shown at the Centennial Exhibition and then at the National Museum had been discovered in Salts Cave. Presumably, by this time Ganter had retired from active cave showmanship, and had also retired Little Alice.

Little Alice had played her role well. Her impersonation of Fawn Hoof had been very successful. Cave visitors were duly impressed, and left satisfied that they had seen the authentic Mammoth Cave mummy. But again, there was no mummy at the cave. When Ganter retired to his farm just east of the old Mammoth Cave tract, he took Little Alice with him. At least part of the time she was stored in his barn. But she was to have new admirers.

George D. Morrison visited Ganter in 1916 to learn all that Ganter could tell him about those underground passages of Mammoth Cave which extended beyond the surface boundaries of the Mammoth Cave lands. Morrison planned to open a new entrance to the cave and show these passages, some of which were grand, extensive, and awe-inspiring. Ganter knew the secrets of the passages and at least one location where an entrance could easily be made. And as with Proctor before him, Ganter now felt no ties of loyalty to the famous cave. He was willing to share his information with Morrison, and he introduced him to Little Alice.

Five years later, Morrison returned to the cave area, opened the new entrance, and developed his eastern half of the cave which he showed commercially under the name, "New Entrance to Mammoth Cave." Bitter rivalry developed between the Natural Entrance management and the New Entrance people.

And adding insult to injury, Morrison produced Little Alice, which he had obtained after Ganter's death, and placed her on exhibition as one of the attractions in his cave. No longer was she simply the Mammoth Cave mummy. Now she had an additional personality. In 1922 he billed her as "The Lady of the Cave. The little girl turned to stone; the most interesting and wonderful of all cave phenomena; a little girl, petrified or mummified by the action of the cave air, a mummy that was found in Salts Cave in 1875; that during the 47 years since the discovery, it had been exhibited in the Smithsonian Institution and at various other places by Mr. H. C. Ganter, the former owner, and now exhibited at the New Entrance. It is believed that the little girl had been captured by Indians, and rather than endure their torture she sacrificed her life."

It was during this period, after she had graduated from the role of the Mammoth Cave mummy to that of the little lady of the cave that she was named "Little Alice." And Little Alice continued to draw attention. After the visitors had concluded their cave tour they were returned to the cave office and shown the mummy of the little girl. Few realized that she was not the original Mammoth Cave mummy made famous over one hundred years before.

Lost John's story began many years ago. He was an Indian, and went into Mammoth Cave in search of gypsum, one of the minerals that form in Mammoth Cave. Chemically, gypsum is calcium sulfate and it is white and brittle. It decorates the walls and ceilings at numerous places. Sometimes it is found as flower-like formations, more often as coatings on the walls like frosting on a cake. Present laws prohibit visitors from taking any as souvenirs. But the Indians over 2,000 years ago were not so restrained; they took large quantities of gypsum for ceremonial or personal use.

Mammoth Cave Indians spent many days in the cave mining this valuable commodity. They explored farther and farther into the branching passages to find new supplies. Chip marks on walls, the remnants of reed torches, and lost moccasins reveal some of the routes they traveled. Unfortunately, many of the artifacts they left were carried away by more recent visitors before effective measures were taken to preserve them.

One day in the fourth century B.C., an Indian miner (probably with others) entered the cave in search of more gypsum. His illumination was a burning torch made by tying long dry reeds together. He carried extra torches and a lunch of hickory nuts. When one torch would burn down he would use it to ignite another.

His woven moccasins stepped lightly on the clay floor and across heaps of broken rocks. At the first fork, he turned to the left and walked past underground springs and immense blocks of limestone that had fallen from the cave walls. Unerringly he chose the correct passages. Gypsum had already been chipped from the walls in these passages. Probably he had made many of the chip marks on various occasions, each time at a greater distance from the entrance.

The passage turned one way, then another. At one place he entered an enormous room so large that his torch did not illuminate the distant walls. Various passages led off this room in different directions. Bearing to the right, he was guided by the sound of a distant waterfall. Just before reaching the place where the water poured from the ceiling and disappeared among the loose rocks below, he took a higher branch to the left. Later he passed through an even larger room than before. All of these walls were bare of gypsum, but some distance beyond he would find it. It had always been so.

He had traveled almost two miles into the cave. There, high on a shelving ledge, was a large detached block of limestone balanced on several smaller supporting stones; on the underside of the large block, the glitter of his reflected torchlight revealed the precious gypsum.

Carefully he worked his way up to the ledge and into the small area under the rock. Using a small piece of limestone picked up in the cave as a hammer, he chipped the gypsum from the undersurface. He, or a companion, dislodged one of the smaller stones supporting the rock above him. The rock shifted.

Desperately he tried to crawl to safety. Only his head, right arm, and shoulder made it. The six-ton rock trapped the rest of his body.

As the months passed the body dried and hardened. Months passed into years and years into centuries. The cave atmosphere changed the body on the ledge into a mummy.

Near the end of the eighteenth century, white men came to Mammoth Cave. Generations of cave visitors followed in the footsteps of the ancient miner, but they were unaware that his mummy was entombed on a ledge twenty feet above them. Through the decades successive guides searched the underground avenues of the aborigines. Other objects of curiosity were found, but the mummy was not discovered until June 7, 1935.

Nineteen thirty-five was a year of transition. Kentucky citizens had purchased Mammoth Cave for a new national park, and the federal government had assumed limited control of the surrounding lands. The cave manager Martin Charlet and his guides—steeped in the folklore and traditions of the cave—were still showing it as they had under private ownership. But changes were being made. Government men were on the scene to evaluate, change, and prepare the area for national park status. Crews of the Civilian Conservation Corps provided labor to make needed improvements. Two veteran guides, Lyman Cutliff and Grover Campbell, led one of these crews almost two miles into the cave to build a safer trail. Cutliff and Campbell were thoroughly familiar with the underground passages; nevertheless, twenty feet above the cave floor on the east wall they saw a ledge which neither of them had examined thoroughly. Carefully they worked their way up the wall to the ledge. Their gasoline lanterns illuminated the sand on the ledge and a narrow passage between the large block of limestone and the cave wall. On their hands and knees they examined the passage and saw, half-buried in the sand at one end of the rock, the head of the mummy. Mammoth Cave had yielded another of its secrets.

Startled, then thrilled, they realized immediately the importance of their discovery. Yet the mere mention of the new mummy would bring hundreds of people to the scene. Instinctively they knew that nothing should be disturbed until experts could be summoned. Cutliff and Campbell reported their find to Charlet, and he notified the National Park Service in Washington. Until experts arrived to take charge, all were to remain silent about the whole affair.

The few who already knew became very secretive. The other guides and cave personnel sensed something important afoot—or rather underfoot. And what could be more important than a new mummy? Hotel clerk Arthur Doyle stepped up to Charlet and two of the guides. Their conversation stopped—but not before Doyle inadvertently learned about the mummy.

"So you finally found lost John?" he asked half in jest. Thus the keen dry humor of the cave region found a name for the new mummy. From that time on it was known as "Lost John."

Archaeologist Alonzo W. Pond of the National Park Service was sent from Virginia. Anthropologist Georg K. Neumann came from Chicago. More tender loving care was given Lost John than all the previous mummies combined. Painstakingly Pond removed the sand from around the head, shoulder, and arm. The skin was a dark grayish-brown color and as hard as leather. Truly this was a remarkable find, undisturbed by amateurs or souvenir collectors.

When public announcement was made photographers and journalists swarmed to the scene. Newspapers gave full coverage, and the Sunday supplements added more. Some readers speculated that this represented a former race which antedated the American Indians, others that he had belonged to one of the lost tribes of Israel. The scientists said very little, pending further study. But for them to continue their work it would be necessary to remove the six-ton block of limestone that covered the rest of the mummy.

The CCC men brought in steel cables and chain hoists. Under the supervision of an engineer these were suspended from solid supports built for the purpose and attached to the rock. Cautiously the rock was lifted and the mummy gently moved to safety.

The scientists were delighted. Although partly crushed and

broken by the weight of the rock, the body was remarkably well preserved. They determined that it was male, approximately forty-five years of age at the time of death, and five feet, three and a half inches tall. A polished mussel shell hung from his neck; a small woven garment was his only clothing. No weapons of any kind were found on or near the body.

All available evidence indicated that the ancient caver, like many others before him, had entered the cave for gypsum. But the experts could only speculate as to why or how they used the mineral. The mummy was placed in a glass case for exhibition in the cave; later the case was moved about a mile back along the main passage to a place more convenient for cave visitors. After methods of radiocarbon dating were developed, tests revealed that Lost John had lived and died over 2,300 years ago.

Lost John had given science new and valuable information about the early inhabitants of America. He was the seventh

"Mummy Ledge" in Mammoth Cave. The six-ton tomb rock trapped the Indian in the cave. His remains became mummified, and are now exhibited as Lost John. The steel cradle and cables by which the rock was raised remain around the rock. *Pete Lindsley.*

mummy found in the caves of the region, and the first to have been so carefully studied. Today he lies in his glass exhibition case in one of the principal passages of Mammoth Cave under the watchful care of the National Park Service. Lost John thus became the Mammoth Cave mummy.

Little Alice did not fare so well. She was placed on a bottom shelf in the old wooden museum off the beaten path, some distance from the cave hotel. She had fewer and fewer visitors, and of those who did come, few were impressed by her present appearance. Her legs were broken and her skin was torn. Truly, she had seen better days.

In the late 1950s a new modern exhibition center was erected to replace the old museum. But Little Alice was hardly suitable for a berth in the new building, and she dropped from sight. Her next abode was a storage warehouse not far from the place where she began her career.

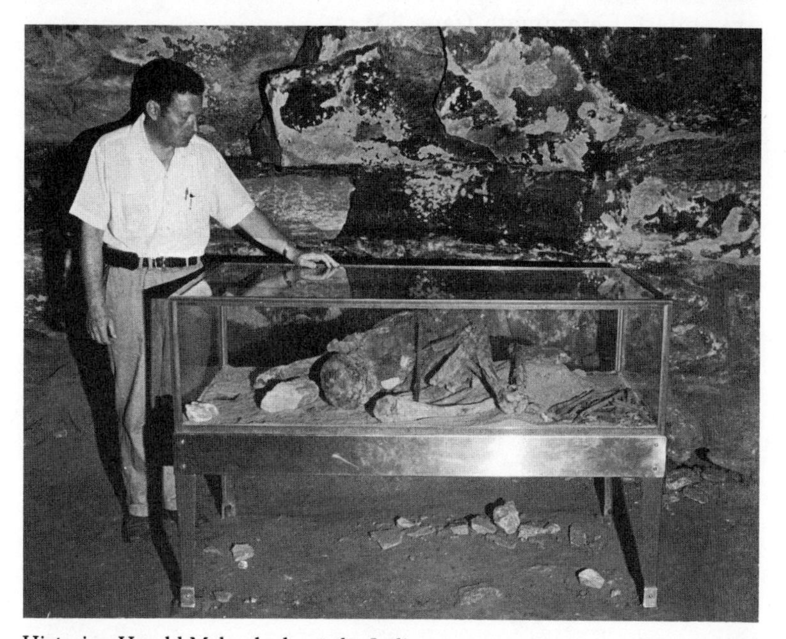

Historian Harold Meloy looks at the Indian mummy, Lost John, discovered in 1935 in Mammoth Cave. This mummy is seen by all cave visitors who enter Mammoth Cave through the natural entrance. *Pete Lindsley.*

Several questions remained unanswered about Little Alice. One question often asked was the age in which she lived. By Morrison's account, she could have lived during pioneer days. When the National Park Service assumed her responsibility, they cautiously suggested she may have been a pre-Columbian Indian. Recently, the Cave Research Foundation has been engaged in scientific research in Salts Cave. Radiocarbon dating of organic materials left by the Salts Cave Indians tells us that she may have lived over nineteen hundred years ago.

But the story of Little Alice is not yet over. The mummy was taken to the University of Kentucky for detailed physical anthropological studies. Preliminary findings disclosed that the body is actually that of a nine-year-old male, who may have lived during the first century A.D. When the studies are completed, scientists will know more about the lives of the early aborigines; then, perhaps, Little Alice can begin a new career as one of the mummies of Mammoth Cave National Park.

The Blue Holes—Underwater Caves of the Bahamas

Cave exploring is a difficult job under the best of conditions. When the cave is underwater the task is even more formidable, requiring the expert skills of both the speleologist and the scuba diver. Specialized and expensive equipment is needed, and it must function without fail underwater and in complete darkness.

The most extensive underwater caves known are the Blue Holes of the Bahamas, off the east coast of Andros Island. The man who knows them best is Dr. George J. Benjamin. Since his first visit in 1957 he has been exploring them at every opportunity, and in his numerous expeditions he has entered more than fifty Blue Holes. A research chemist, Dr. Benjamin operates a film-processing laboratory in Toronto, Canada. He has brought his expertise in photography underwater, designing many of the lights, cameras, and procedures himself. His photographs, films, and stories have appeared in *National Geographic* Magazine, on Jacques Cousteau's television show, and in many other places. His film, "Andros Blue Holes," was named *Best Documentary of 1969* by the Photographic Society of America.

The Blue Holes—Underwater Caves of the Bahamas

GEORGE J. BENJAMIN

Much of my life has been spent climbing mountains and exploring caves. Both provide adventure and physical challenge. Both appeal to the explorer's desire to walk where no man has walked before. But there are differences. The mountaineer always has his target before him. The sight of the summit helps generate strength for the final assault. The caver can only grope in the dark. He has to endure the disappointments of dead ends and crawlways too small to fit through. What keeps him going is the hope that one day he will break through into unexplored territory.

For me, there are added thrills when the cave is underwater. Such underwater caves are found off the east coast of Andros Island in the Bahamas. Seen from the air, the dark blue patches of

---

Three divers descend into a near-vertical Blue Hole. Light from the sky outlines the entrance. *George J. Benjamin.*

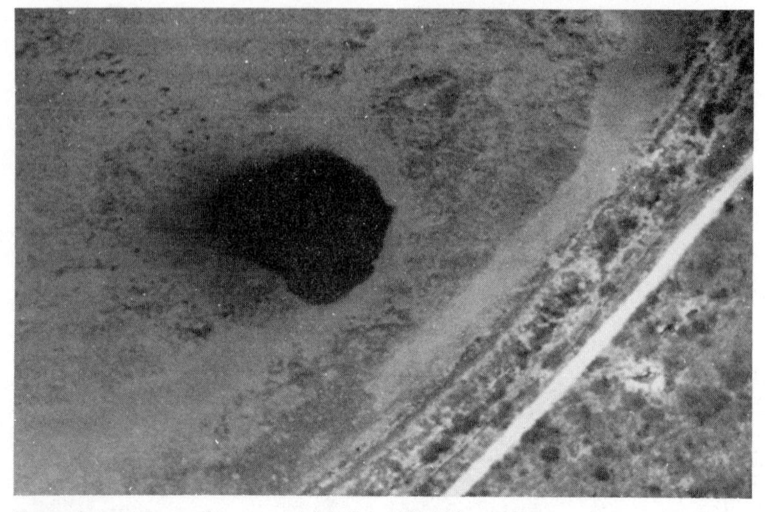

From the air the underwater sinkholes called Blue Holes can be spotted easily. Most, like this one, are close to shore. *George J. Benjamin.*

A small whirlpool forms over this Blue Hole as the tide rushes in. When the tide changes the current reverses and water flows out of the cave. The diver waits for the few minutes of slack water between tides to enter the cave. *George J. Benjamin.*

the underwater cave entrances contrast sharply with the light green water of the lagoon. These are called the Blue Holes. Since 1960 I have used all available time away from my laboratory in Toronto to explore them.

Submarine springs have been known in many parts of the world since ancient times. The Blue Holes are distinctive because of their strong reversing currents, which follow roughly the pattern of the tides. For six hours crystal clear salt water runs out of the cave as though from a gigantic spring. Then, with the rising tide, the water from the lagoon is sucked back in with tremendous force. This great in-rushing of water often forms whirlpools.

The Bahama Islands sit at the top of the tallest chain of limestone mountains in the world. Most of it is underwater. The peaks of this huge mountain of limestone are barely above sea level, and make up the present-day islands. Offshore, underwater cliffs drop steeply to the ocean bottom—over 27,000 feet down. The reef building which formed the islands began in Cretaceous times, some 120 million years ago, and has continued with minor interruptions until the present. During Pleistocene time much of the surface water of the earth was frozen into glacial ice. The ultimate source for the ice was the ocean; during glacial times the ocean levels were several hundred feet lower than at present. Caves formed by solution during these periods of lowered sea level when lenses of fresh water above the level of the ocean occupied the present positions of the caves. There were several ice ages with warmer times in between. As ocean levels moved up and down to correspond with the growth and melting of the glaciers, caves would be submerged by ocean water, filled with fresh water, or exposed to the air. Finally some 10,000 to 15,000 years ago the glaciers melted for the last time. Sea levels rose to present levels and the caves were flooded. These flooded caves in the Bahamas are known as the "Blue Holes."

With the help of my two sons, Peter and George, Jr., I have explored fifty-four of these submarine caves since 1957. The most important and the largest Blue Hole we have found so far is located in the south bight of Andros Island. It took five years to explore its unique features, and there are still many more passages unexplored.

I spotted the area from the air many years ago; but it was not until Christmas 1967 that Peter and I had a chance to explore it. As soon as we entered we realized that it was much larger than anything we had ever seen before. We proceeded cautiously,

The divers explore some of the steep underwater cliffs. Some cave entrances open directly on the face of the cliffs. *George J. Benjamin.*

step by step. Sixty feet in there was a vertical pit. The sight down was spine-tingling. The cave disappeared into blackness beyond the power of our searchlights. We belayed our vital lifeline and proceeded downward. It took us three nervewracking dives to reach the bottom at 160 feet. There the cave had only just begun.

In the summer of 1970, scientist-explorer Jacques Cousteau came to Andros Island in the research ship *Calypso* to make a special television documentary. I was asked to lead the expedition into the depths of the Blue Holes and to outline the mysteries we had uncovered. I worked with Dr. Robert Dill, the geologist of the project. He had one crucial question for which we had no answer. The Blue Holes, as we know, formed as normal land caves. Most had once been dry and filled with air.

The divers did not know if this was the remains of an actual stalactite, or simply residue from the original reef material. Testing indicated it was not a cave formation. *George J. Benjamin.*

But no stalactites, stalagmites, or other speleothems had ever been found. Why should this be so?

We could think of at least two explanations. First, during times of low water the climate in this particular area was exceptionally dry. Speleothems are built by dripping water which seeps from the surface into the cave. Perhaps there simply was not enough rain. If the arid climate theory is true, then there are no speleothems and it is hopeless to search for them. Another theory postulates that originally there were plenty of speleothems. Later, as the glaciers melted and the ocean rose, the whole area was flooded with salt water. The chemical action of the seawater changed the original material, so that only the outer forms of the speleothems were preserved.

I took Dr. Dill to a nearby Blue Hole containing formations resembling stalagmites, but the results were inconclusive. It

Despite the warm water the divers usually wore long pants to avoid cuts on the sharp rock. Hanging below the diver is a battery-operated diver propulsion vehicle, although it is not in operation in this narrow passage. The dark object on the side of the vehicle is a strobe unit for taking photographs. *George J. Benjamin.*

could just as well be residue from pillar coral or even beach rock deposited when the reef was formed. Since the original reef material is amorphous and not crystalline, it would be necessary to find a crystalline core, one showing that it was originally deposited by dripping water.

We made many dives searching for such formations. I took Cousteau's team into every cave I could. We found many would-be stalagmites but no crystalline cores. Finally, the *Calypso* had to depart. But I was left with a determination to go on, to go deeper and further in our quest.

By this time, we had reached the limit of our technical ability and had explored an area of 300 feet around the main pit. Before further exploration was possible we had to adapt our equipment and methods for deeper dives and penetration. Safety was the prime consideration. The breathing apparatus, the lighting, and the auxiliary systems could have no flaws. The size and weight of all equipment had to be kept to a minimum and still provide safety. Plastic reels were designed to eliminate bulky coils of safety line. We fixed over a mile of permanent lines as guides in and out of the caves. By tabulating the tides and the time lags we could extend our safe diving time, entering and leaving at slack water. We could not chance being trapped in the cave by a tidal whirlpool.

With these improvements we penetrated 1,000 feet inside the cave at an underwater depth of 260 feet. Some of the pits were so deep we could not see the bottom. The cave was so big that in our search we had to neglect numerous smaller openings. When we did find time to explore them, we discovered two pillars, three feet tall, which looked suspiciously like real cave stalagmites. Were they genuine, or simply residue from pillar coral or beach rock? We had been disappointed many times before, but our scientific approach required that we follow up every possibility. Naturally we did not consider breaking off the formations. We would have to return with a core drill, to take a small sample. As we left, I looked back. High in the roof I saw a solid column formed by a stalactite joining a stalagmite. I felt, beyond doubt, that it was a real speleothem.

We went on. Eight hundred feet into the side passage we

came to a huge room which we later named the Grotto. We were at the end of the lifeline and had to stop. The clarity of the water was superb, and within the reach of our light I saw the faint outline of a cavern filled with stalactites and stalagmites. I switched off my own lights, sat a moment in total darkness, and then lit up the cavern by firing the strobe. By the light of the brilliant flash I could see that this was only the beginning of a vast network of passages. But our time was running out. Minutes

Cautiously the diver traverses a passage far below the surface, shining his light ahead as he goes. *George J. Benjamin.*

had passed quickly, and our air supply was running low. We had to hurry back.

This brief glimpse of the Grotto opened up a new chapter in underwater caving. I contacted Dr. Dill, so that our discovery could be included in Cousteau's film. When we returned, the weather was windy and visibility in the cave was so poor that we could not find the hidden passage. On a later trip, we laid a permanent lifeline into the room. We pushed on fifty feet further and entered a maze of passages between ten-foot-tall cave formations.

The passage continued. We descended a pit 260 feet without seeing the bottom. A high, level passage continued beyond the pit; we followed it for 600 feet without finding an end. Later, with Cousteau and his divers we used underwater flares to

This ten-foot-high stalagmite could only have formed when the cave was drained and filled with air. *George J. Benjamin.*

illuminate this cavern. I watched from the same position where only a few months ago we had seen the Grotto for the first time. Then there had been only a faint light; now there was a galaxy of burning flares.

With Dr. Dill we went to the Grotto to look at speleothem samples. Most of them were brittle, honeycombed, and partially destroyed by salt water. We continued our search. Finally we found a stalagmite with what looked like a crystalline core. I fired my strobe close by. The crystal glowed for a fraction of a second with a pale green light. It was a true speleothem of crystalline calcite. So the Grotto was found and the stalagmites identified, proving that these caves had once been filled with air. The caves must therefore have been formed before the last glaciation, at least 10,000 to 15,000 years ago.

Inland Sinkholes

The source of the great volume of water exhausted from the Blue Holes has puzzled many people. It pours out in great volume, millions and millions of gallons. But where does it come from? The most plausible explanation is that during rising tide seawater is forced into the Blue Holes and distributed underneath the island through underground channels or by seepage through porous rock. During low tide the pressure is released and the direction of the flow reverses. In simple terms, the island water table acts like a piston in a supersize cylinder, changing direction with the tides. Fluctuation of the level in inland pools has actually been measured.

The existence of perfectly round ponds scattered throughout Andros Island is well known. Aerial surveys show more than one hundred. They are similar to sinkholes all over the world, but these are filled with water. Since airtanks and gear weighing up to one hundred pounds per diver must be carried in, only a very few can be considered within walking distance of roads or creeks. The terrain is rough, the brush is dense, and there are poisonous plants. In the sinkholes rainwater sometimes accumulates on top of the salt water with a minimum of mixing, forming a freshwater lens. In many sinkholes the fresh water is crystal clear; submerged divers can study the surface features clearly from underwater.

In fresh water, which has less buoyancy, divers sink rapidly until they hit salt water twenty to one hundred feet down. There they bounce up, as if from an invisible trampoline. The separation of zones may be sharp or gradual; the salt water is dark brown and murky from organic matter. Sometimes after a heavy rain there is such a concentration of organic matter that fish and other life are killed.

The water clears up as one dives deeper, but it is dark and lights must be used. There are many overhangs and openings to

The divers adjust their eyes to the dimness a few feet under the surface as they get ready to explore. Ripples in the surface a few feet over their heads can be seen clearly. *George J. Benjamin.*

passages, but in the dark brown water exploration is difficult. Also, as the clarity improves and the bottom is approached the water becomes warmer—probably because of the absorption of the sun's infrared rays, which pass freely through the upper salt water.

In most of the sinkholes the upper part has steep vertical

The divers found many submerged cave formations. *George J. Benjamin.*

walls, like a huge barrel. Farther down there are overhangs which may lead to a cave system. Opposite the overhangs are muddy slopes stretching downwards to the greatest depths.

In 1970 the British Hydrological Survey began exploration of the freshwater resources of Andros Island. Sinkholes provided an easy access to the saltwater table. Since none of the B.H.S. personnel was a diver, we were asked to do the underwater spelunking. By helicopter we surveyed more than thirty sinkholes, and were able to dive in eleven of them.

Our most important discoveries were several huge stalactites in a sinkhole near the center of the island. The sinkhole was more than 300 feet across. About eighty feet down there was an overhang with at least ten stalactites more than twenty feet long. One of them measured twenty-seven feet in length. We named

The divers swim very carefully to avoid hitting the numerous stalactites and stalagmites revealed by their lights. *George J. Benjamin.*

this sinkhole Archie's Cave, in honor of our former diving companion, the late Archie Forfar.

In several sinkholes we found what looked like sodastraw stalactites: they were about three feet long, soft and chalky looking, with holes through the middle. Some speleologists believe that growth of sodastraws is an essential stage in the formation of stalactites.

In other holes we found stalactites formed around what appeared to be roots sticking through the roof. But none of these speleothems looked exactly like formations found in caves filled with air. Perhaps they formed by a different process, maybe through biochemical action. In places stalactites were encountered in great numbers. Most of them ended abruptly at the freshwater-saltwater interface and did not extend down into the salt water. Some broken fragments were found lying in the mud at the bottom of some of the sinkholes.

Our secret hope in all these dives was to find a connection between the island sinkholes and the Blue Holes in the ocean. But we found no passages branching out from the more or less vertical shafts. Our explorations were hindered by the dark and muddy waters. There is a native fable of a duck that disappeared into a sinkhole. A week later it popped up in a Blue Hole—alive, but with all its feathers missing. The average Blue Hole pours out some 400 cubic feet of water per second. The usual Andros Island sinkhole covers an area of roughly one acre, about 40,000 square feet. If it were directly connected to the sea, the water level in the sinkhole would rise or fall a foot every two minutes. The actual fluctuation during the tidal cycle is only three to four inches. Probably the water filters through tiny channels too small for a man. Maybe they are duck size. We didn't find a passage to the ocean. But there are other trips, and other sinkholes. We'll be back.

Histoplasmosis: Cave Explorers' Pneumonia

The environment in caves is generally healthy. Despite some popular beliefs, the air in caves is usually clear and fresh, and the caves are well ventilated. Most disease-causing organisms cannot live in the dark, even-temperatured, moist environment. Accidents occasionally happen, but these are most often caused by people, and not by some natural calamity.

But in some areas stories were told of caves that made people sick. Some caves were reputed to have buried treasure, and those who sought it would become ill—perhaps as punishment or part of a curse. Entire groups were known that became seriously ill after visiting certain caves.

Warren C. Lewis, M.D., a physician in private practice in the cave country of Illinois, likes to write about both his hobby, caves, and his profession, medicine. He combines the two as he unravels the story of the mysterious "cave sickness."

Histoplasmosis: Cave Explorers' Pneumonia

WARREN C. LEWIS, M.D.

Rumors of buried treasure swept through the town of Foreman, Arkansas, one fall day in 1947. Jesse James, the notorious bank robber and outlaw, the story went, had buried his loot in a local cave. This cave had just been rediscovered in a rocky outcrop outside of town by a visiting couple from Oklahoma City who were digging there. They even had a bulldozer. Their story suggested that De Soto or some other early adventurer might have used the cave as a hiding place for gold. In the White Cliffs area they uncovered the top of a flight of steps leading into an underground opening. This led to a steep dirt-filled stairway hewn from solid limestone into a hidden chamber.

Once they had cleared the entrance they found themselves in a good-sized underground room with substantial rock walls and ceilings. From this cavern low crawlways led off to several smaller rooms. The air was lifeless and dusty. The cave was piled high with dirt, broken rock, and organic debris. The discoverers

---

This cave is located near the lower levels of Grand Canyon in Arizona. Histoplasmosis has infected visitors to several caves in the area. *Gene Griffen.*

found that ventilation was extremely poor, and they stirred up clouds of dust. Nothing of value was found. Finally, in disgust, they gave up their search for the elusive hoard and went back to Foreman where they told of their cave discovery. Shortly afterward they returned to Oklahoma City, where they were soon hospitalized with a mysterious respiratory illness.

Others came to search for the reputed treasure. Twenty-five adults and teenagers crowded into the cave one September day in 1947. Armed with flashlights and hand tools, they went to work. After searching each rocky crevice, they dug into the cave floors with vigor, moving mountains of dirt and raising clouds of irritating dust. Some of the more sensitive ones choked on the dust and could hardly get their breath. No treasure was found, but four days later several of the explorers got sick. The illness started with a runny nose and stuffy head cold, and the victims felt weak and listless. In about twelve hours many got a severe chill with high fever, chest congestion, and chest pain. Most found it difficult to breathe. By the thirteenth day nearly everyone who had been in the cave had become ill. No one was infected except the cave explorers. Twenty-one local people were sick with what seemed to be a peculiar kind of cave-caused pneumonia.

A husky twenty-three-year-old planter was one of those who had worked energetically in the cave. On the fourth day he developed symptoms of a cold in his head and chest. By evening his temperature had reached 105 degrees Fahrenheit, he was sweating profusely, and became extremely irritable and jumpy. He could not tolerate noise or bright lights, and was almost delirious. The acute stage of the illness lasted several days, and then his fever broke. He became lethargic and lapsed into a stupor for several more days. On the seventh day he was so weak that he was barely able to get out of bed. He had lost twenty-four pounds. For the next several months he was weak, and short of breath on exertion.

Striking changes were observed on his chest X-ray. It appeared as if a white snowstorm had obliterated the details of the normally dark lung areas. The lungs were studded from top to bottom with a multitude of tiny inflammatory granules. The films resembled X-ray pictures of miliary tuberculosis, a rare form of rapidly spreading tuberculous disease. The diagnosis was "acute miliary pneumonitis."

His illness did not respond to either penicillin or the sulfa drugs, and his cough was not relieved by ordinary expectorants. However, his headache was temporarily relieved by codeine. Even though he was extremely ill, no one nursing him or caring for the other victims came down with the sickness. In fact, no one in town who was not in the cave came down with the fever. It seemed obvious to the local citizens that the cave air was poisonous to human beings. A bulldozer was used to widen the entrance and to scrape away part of the face of the cliff. The original entrance was destroyed and the main underground room was opened to light and air. This also made it less attractive to children in the area.

The Arkansas State Board of Health sent a team of trained epidemiologists to Foreman with a mobile field laboratory and a doctor from the U.S. Public Health Service. They tested the cave air, analyzed the cave rock, and examined many of the victims. They were able to rule out a score of possible causes, including gas inhalation, malaria, dengue fever, influenza, relapsing fever, and silicosis. They questioned the possibility of the victims having virus pneumonia or Q-fever.

The examiners suspected the disease may have been caused by a fungus or mold. Many such organisms thrive in dusty soil such as was present in the cave. But skin tests of the patients with diluted extracts showed nothing, and other blood studies were also negative. The medical team, after a comprehensive study, reported that the cause of the disease was not known.

The victims gradually got better. However, the U.S. Public Health Service team did not give up. They returned several times over the ensuing years to run more tests, often with improved or more sensitive materials. Surprisingly, they found that nearly every victim—when tested months later—had a positive skin test for the fungus *Histoplasma capsulatum*. This fungus, at that time known to occur occasionally in bird droppings, can cause a pneumonia-like severe lung disease called histoplasmosis. Other blood tests were not inconsistent with such findings. Was histoplasmosis the mysterious cave sickness?

The next step was to test the cave soil for the pathogenic (disease-causing) fungus. No one had done this successfully before. Ordinary culture methods had yielded only harmless bacteria or fungi. But within two years Dr. Chester W. Emmons perfected a method for testing soil that involved making a suspension of the dirt in a saline solution. This was then injected into mice along with antibiotics to suppress any bacterial infection. After several weeks the mice were sacrificed and cultures made from internal organs.

Finally dirt from the Foreman cave was tested. There, growing in the cultures, were the fateful colonies of *Histoplasma capsulatum*. The epidemic was shown to be histoplasmosis.

Histoplasma capsulatum is a dimorphic pathogenic fungus. One form is that of a harmless mold that grows slowly in protected areas in moist fertile soils. There it forms a branching network with occasional spore-bearing branches or hyphae. Several types of spores are formed. An established colony produces millions of spores that are spread by the wind. If a spore lands in a suitable place it starts a new colony like the one it came from.

But if the spore is inhaled by a warm-blooded creature, it is transformed into a corrosive swollen yeast cell that reproduces by budding. Although related to the common yeast cell from which we get bread and alcohol, this yeast cell acts quite differently. Inside the body it eats its way through the alveolar or inner wall of the lung, causing serious inflammation in susceptible individuals. It may be carried in the bloodstream to distant organs such as the spleen and cause inflammation there. Multiplication of cells may continue until the animal is overwhelmed. It is the yeast form that causes the disease and is found in the tissues of victims of histoplasmosis.

In 1958 three members of the Mid-Mississippi Valley Grotto of the National Speleological Society visited Bat Cave in Pulaski County, Missouri. They were exploring crawlways in the upstream passage, hoping to find passages that had not previously been discovered. The only way to go was to crawl over rock slabs covered with bat guano. On the way out of the cave they stopped to rest. One caver remembered that his rose bushes at home were not thriving. He filled a plastic bag with guano and his companion did likewise.

A few weeks later the rose grower was hospitalized with a severe respiratory infection which proved to be histoplasmosis. In spite of intensive medical treatment in the hospital, his condition gradually became worse. The disease had spread beyond the lungs and had involved other organs. Eventually, one lung was destroyed, and indirectly he lost one kidney. He underwent a long period of hospitalization and suffered permanent impairment of his health. His companion did not get sick. However, when his skin was tested later for sensitivity to histoplasmosis, the entire arm turned red and swelled up to the elbow.

An evening bat flight at Carlsbad Cavern, New Mexico. Most of these are Mexican freetail bats. Large colonies of bats live in several southwest caves, and some have been known to carry histoplasmosis. *National Park Service.*

At about the same time three other members of the MMV Grotto explored another Bat Cave, in Shannon County in central Missouri. This cave also contained piles of bat guano. Some of the back passages were so low that the cavers had to crawl on their hands and knees. One of the group, Barb MacLeod, came down afterward with histoplasmosis. Fortunately she had no complications, and recovered.

Two other epidemics in the Oklahoma-Missouri region had already been studied by the Army Medical Corps and the U.S. Public Health Service. In 1944 two groups of enlisted men at Camp Gruber, Oklahoma, were affected. During a storm they had sought shelter in an abandoned cellar while on a training exercise. Most of those who were in the cellar came down with an unidentified illness. Similar cases were observed among servicemen at Camp Crowder, Missouri. They had been assigned to clean up or demolish abandoned farm buildings in the camp area. Soil samples taken several years later from these two sites were positive for *H. capsulatum*.

If there was any doubt about the presence of histoplasmosis in Arkansas and Missouri, it disappeared when skin testing became widespread. A broad histoplasmosis belt lay across the center of the country around the Ohio and Mississippi valleys, with Kentucky and Tennessee in the middle. An extension lay along the Appalachian Mountains to Pennsylvania; another extended southwest along the Mexican border states. This included many of the primary cave-forming areas within the borders of the contiguous United States.

The development of a simple skin test enabled investigators to identify persons who had already had the disease. Eighty percent of the adults in Arkansas, Missouri, Kentucky, and Tennessee were found to have positive histoplasmosis skin tests. These persons were relatively immune to the disease. Only a small percentage of those living far away on the fringe of the area showed positive reactions.

Many epidemiologists think that the disease originally came from moist tropical lowlands. At present the fungus has a worldwide distribution in tropical countries. In these areas, such as Panama, everyone is exposed during childhood and the fungus is widely prevalent. But in temperate latitudes it thrives only in protected moist locations in which competition from bacteria is not overwhelming. It grows best in soils enriched by bat or bird guano with a high phosphate and nitrogen content. It will grow under buildings or woodpiles, in hollow trees or chimneys, or under bird roosts; but it grows especially well in caves.

The underground climate in most caves is steady and unchanging, moist and undisturbed. Deposits of clay fill are common in caves, and these may prove ideal for fungal growth. At least part of the year the temperature must reach 70 degrees Fahrenheit for the fungus to grow. The Ozark plateau and the Ohio River valley of the United States provide these conditions in many locations.

Histoplasmosis was once called the "Ohio Valley disease" because most of the positive reactors in this country came from that area. These central states are rich in caves that can provide a suitable microclimate. Missouri leads the nation in caves: some 3,000 are known and the number is growing. Tennessee has over 2,000 caves, mostly in the east and central portions of the state. Kentucky has more than two hundred miles of passage in the Mammoth-Flint Ridge area, plus many other caves, some of which harbor large bat colonies. Arkansas also has many extensive well-decorated caves.

It appears that bats, like other animals, contract histoplasmosis by inhaling infectious particles that then lodge in the lungs. When bats cling together in large clusters the disease can be transferred easily from one to another. Studies in Florida indicate that bats living in groups are more prone to this infection than solitary bats. However, even bats of different species or tree-dwelling bats may be infected occasionally, apparently by sharing the same cave or roost. It is not uncommon for several species of bats to be found in the same cave. They may share transitory resting caves on long overland flights, or occupy the same daytime caves. Bats most often gather together in large groups in caves for winter hibernation. Sometimes they congregate from widespread areas to winter in suitable caves. Some species roost near the entrance where the winter temperature is often low; others prefer the warmth farther inside the cave. Many prefer an area of high humidity. If such caves are infected with histoplasmosis, many bats may be infected. The fungus, of course, cannot grow in the wintertime; but it will remain dormant, and the spores can be picked up by any passing creature if the cave dirt is disturbed.

If the disease originated in the tropical lowlands of Central or South America, it may have been carried by bats to the central Mexican highlands. Migratory bats then could have transported the disease to the huge limestone caves of the Southwest and the gypsum caves of Oklahoma. Bats from eastern Oklahoma often winter on the Ozark plateau. From the Ozarks bats migrate in all directions in the spring. Kentucky bats fan out over the north-central states. Florida caves may have been infected by flights from islands in the Caribbean. All these may be links in the chain of spread of histoplasmosis.

However, relatively few people get histoplasmosis from caves. Many more get it from activities related to the roosts of birds. Victims in Wisconsin got the disease from digging along a hedge for angleworms, from digging shrubbery around a church, and while building a house on a lot once used as a bird roost. In the last instance, many people got sick during construction: the bulldozer operator, the men who set the footings, and the plumbers. Carpenters from a nearby roofing job got sick after they came over on their break to watch the footings being poured. The city inspector became ill. Even the health officer who investigated the epidemic developed a lung lesion and had to have chest surgery. A farm family in Indiana became infected while cleaning out a silo; people were infected in New York State from a hollow tree, and in Maryland from cleaning up a dusty house. A bird watcher caught the illness while observing prairie chickens on their booming grounds in Wisconsin, and other birders were infected while studying a gull rookery on the shores of Lake Huron. Histoplasmosis is far from rare. It is estimated that forty-six million Americans have been infected. It is asserted that histoplasmosis is the most common systemic fungus disease. But most people in infected areas have very mild cases, and then develop immunity.

In January 1973 a group of teenagers attending a church-sponsored camp in Inwannee County, Florida, explored a limestone cave housing numerous bats. Several other Florida caves were previously known to be infected. Twenty-nine campers explored the cave. Part of the group went in twice for about thirty minutes each. While in the cave some members of the party amused themselves by throwing dirt at the hanging bats in order to make them fly. This created a dusty atmosphere.

A few days later an eighteen-year-old girl from the group was admitted to the University of Florida Medical Center with fever and severe chest pain. She was unable to get her breath. The cause of her illness was not clear to the staff. On her third hospital day the doctors learned from her mother that three of her companions had been hospitalized with similar symptoms. They all had a cough, fever, night-time sweating, shortness of breath when they moved about, and chest congestion. In all, twenty-three persons from this group had what proved to be acute pulmonary histoplasmosis.

Florida has been under the scrutiny of public health officials ever since cases were reported from the northern counties several years ago. Infected bats and caves have been found in seven counties in north-central Florida. However, the people living in Florida have a low rate of reaction on skin testing. Most have had little contact with wild birds, bats, or caves. But nearly all members of a group of Florida cave explorers were strongly positive.

More reactors are found in Georgia than in Florida, but the distribution is quite uneven. The sources of infection are not fully known. In one study numerous infected bats were found in the attic of a schoolhouse. Many Georgia animals have been found to have positive histoplasmin tests. These include the common domestic animals as well as the Norway and brown rat, striped and spotted skunks, gray and red foxes, opossums and raccoons. When dogs get the disease they usually have congested lungs and diarrhea. Seriously ill dogs succumb to the illness in one or two weeks in spite of intensive treatment. Any carnivore may get the illness from eating an infected animal. However, it appears that most animals get histoplasmosis from inhaling infected dust in dens shared with other animals. Grazing animals can get it from sniffing infected soils.

Crabtree Cave, the longest cave in Maryland, is formed by a series of narrow fissure passages. Even a thin caver has to squeeze, shinny, and crawl. Perhaps it is fortunate that the cave is so difficult, because several cases of histoplasmosis have been noted from cave trips there.

Maquoketa Cave State Park in Jackson County, Iowa, has a charming old-fashioned picnic area. It boasts an electrically lighted cave complete with concrete walks, stairs, a natural bridge, and a flowing stream. There are picnic and camping areas and a refreshment stand. A series of shelter caves along the creek provide adventurous settings for children. No one can get lost in them and lights are unnecessary. The park has been popular with several generations of Iowa youngsters.

A survey of 3,146 Jackson County schoolchildren for histoplasmosis in 1955 showed a very uneven distribution of positive reactors within the county. In a single family some children would be positive and others would not. This suggested that the children were being infected away from home rather than on their own farms. By far the greater number of positive reactors were younger children in the western part of the county. Most of these children lived just outside the state park. Two soil samples taken at a cave entrance in the park were positive.

Radar technicians at Edwards Air Force Base in Texas reported a most unusual pattern one evening. A twisting plume of swirling creatures rose into the atmosphere, gradually spreading out over several hundred miles of Texas sky. It was a flight of millions of bats from nearby Bracken Cave. But they were not dispersing to their regular feeding grounds along the lush green river valleys. The weather map provided an explanation. A mass of cold air had pushed in from the north, causing a thermal inversion. Heavy cold air was sliding under the warm air near the ground and lifting it several thousand feet. The flying insects were in the high, warm layers and the bats went up to get them. It was hours before these denizens of the Texas caves retraced their aerial path to their underground retreat and hours before those particular air lanes were safe for plane travel.

Most of these were Mexican freetail bats which live in colonies in the Southwest. Large colonies are found in thirteen caves in Texas, five in Oklahoma, and isolated caverns in New Mexico, Arizona, and Nevada. They are highly beneficial to man, since they feed almost entirely on flying insects. The guano produced by them is a valuable resource, and it is bound to become even more valuable if artificial fertilizer becomes scarcer. In Frio Cave, Texas, a railway has been built to haul out the guano. The activity can only be carried out in winter when few bats are present and cave insects are at a low ebb.

Histoplasmosis was first isolated in Texas from a bat in Frio Cave. Later the fungus was isolated from soils in other Texas caves. The bat caves of Texas are distributed along a line from Lampasas County to Uvalde and Edwards counties, a distance of 250 miles. Ten of these caves harbor summer colonies of a million or more bats. Bracken Cave has been worked commercially for sixty years, producing about eighty tons of rich fertilizer a year. Bat Cave Cavern and Ney Cave have also been mined.

The fungus has been found in guano at Carlsbad Cavern and in infected bats in this and neighboring caves. A government epidemiologist was reinfected during a visit there. He was known to have a positive skin test before visiting the area. In spite of this he had a typical case of mild pulmonary histoplasmosis after his visit. A state archaeologist working in New Mexico to preserve Indian artifacts also became sick with histoplasmosis. Many bats in New Mexico, Arizona, and Nevada use abandoned mines as roosts to escape the heat of the sun. Members of two families were infected in an old mine hear Hobbs, New Mexico. In the past few years several cavers have caught histoplasmosis in Crocketts Cave, New Mexico. Several miners harvesting guano in a cave in a side canyon off Grand Canyon in Arizona became ill.

Bats appear to play a unique role in the long-distance spread of histoplasmosis. Other animals may also play their part in its distribution. The fungus has been recovered from the feces of dogs. It has been found in the nasal passages of infected domestic rabbits and has been recovered from the air in their cages. It has been found in a dog tick that had fed on an infected dog. It has even been found on feathers in a feather pillow. But none of these sources of infection seems to be very important compared to bats.

A few infected bats will develop ulcerations in the wall of the bowel in the intestinal lymph glands. They will discharge the fungus from these sores into the intestinal canal to be passed in the feces. By flying from one location to another the sick bat may infect a new site many miles away. This site may be a cave, the loft of a building, or the soil below the shutters of a church. Bats frequently defecate while roosting. Cave explorers who have been under a large roost do not forget the fall of pellets like gentle rain, or the pinging sound as droppings bounce off hard hats. Bats also defecate frequently while feeding—observations indicate as often as every fifteen minutes—so the droppings may fall anywhere the bat flies. If infected feces land in a moist protected spot in enriched soil, the fungus can grow. Once established in a suitable location, the fungus can continue to grow for a long time and over a large area. Infected soils have been found in very dry caves and in open fields long after the woods had been cleared from the land. Perhaps the hardest thing to understand is why the fungus is not found more often in routine soil samples. In spite of widespread infections of humans and animals, the infected soils remain rare and are usually restricted to confined spaces.

Within the United States conditions favorable for the propagation of *Histoplasma capsulatum* exist virtually from coast to coast and from border to border. From 1957 to 1966, 746 deaths from histoplasmosis were reported. The actual figure may be higher. Ninety-two percent of the epidemic outbreaks in the United States have been associated with bird habitats. However, some of these sites may also be used by bats as well as birds, since bat roosting sites are often overlooked by the general public unless they are specifically sought out.

Histoplasmosis is widely distributed in Central and South America. Some states in Mexico have many positive reactors. Venezuela, Panama, and Guatemala are also high. Students from the mainland at a center for cave studies in Puerto Rico have been plagued by the illness.

A histoplasmosis vaccine would be very useful for those in high-risk groups, which certainly includes many cave explorers. No vaccine for humans is yet available, although rabbits can be completely protected by an injection of heat-killed spores. In histoplasmosis, immunity is everything. It should not be too difficult to make a human vaccine to provide adequate immunization.

The Lava Caves of Saddle Butte

Lava caves or tubes may form in volcanic areas as molten rock, or lava, flows from the depths of the earth onto the surface. Since they form only a few feet directly below the surface of the flowing lava, their existence is limited: if the tube is not filled or covered by a later flow, erosion and collapse destroy them in a few thousand years.

There are many lava caves in the western United States where lava flows have occurred recently, some within historic times, and new caves have been found in fresh flows in Hawaii. Idaho, Washington, Oregon, and northern California have large numbers of such caves. One of the most extensive lava cave systems is in the Saddle Butte area of southeastern Oregon, where more than a dozen caves developed in a single lava flow. The caves provide shelter for many animals, and although they have been known and were used by settlers in the late 1800s, they were forgotten until rediscovered in the 1960s.

Charlie Larson and his wife, Jo, are professional photographers in Vancouver, Washington, who combine the best of their interests to produce outstanding cave photographs. Both are Fellows of the National Speleological Society. Larson, editor of the *NSS News* for many years, is currently president of the National Speleological Society.

The Lava Caves of Saddle Butte

CHARLIE LARSON

Lava caves or lava tubes are the product of molten basaltic rock, gravity, and an inclined surface. Only relatively fluid lavas, known by the Hawaiian term *pahoehoe*, produce long caves, and then only on gentle slopes under favorable conditions. Large lava cave systems form in long lava flows whose length, parallel to the direction of flow, is several times the width. Such flows occur when the highly viscous but still fluid lava is vented onto an inclined plane or follows a pre-flow topographic low such as a streambed or valley. Heat loss, turbidity, and friction immobilize the flow's top and outer edges, creating hardened margins or levees which channel fresh lava from the vent to the advancing front or toe of the flow. Under favorable conditions the channel of still-moving lava will reduce to a width which can be spanned by a cooled and hardened top or self-supporting bridge of lava, and a roof is formed. Such roofs may be strengthened by cooling, accretion of additional layers of hard-

---

The lower end of Baker Cave floods occasionally, as shown by the mud floor and cracks, and by the waterline halfway up the wall. *Charlie and Jo Larson.*

ening lava within the channel, additional overflow layers on top, or all of these.

Lava under the surface of the flow may still be molten, even though the roof has hardened. If the last flow of lava to occupy the roofed-over channel drains away, it leaves a void—and a lava cave is created. Because of their relatively simple passage plan and appearance they are also called lava tubes. Slope is a critical factor in the formation of lava tubes: too steep, and the lava flows

Baker Cave, longest of the Saddle Butte system of lava tube caverns. Light from the entrance can be seen in the distance. As the lava drained from the tube, the cooling but semi-fluid walls bulged into the void. *Charlie and Jo Larson.*

too quickly for roofed-over channels to form; too little slope and the lava spreads out and cools as a mass.

Lava caves don't have much in common with limestone caves, except inner darkness. Their origin is as complex as that of limestone caves, but less is known about them since they haven't been studied so extensively. Lava caves are the unusual product of many variables, including the lava's chemical makeup and temperature, its gas content, lava flow volume, and topography.

Lava caves develop just below the surface of the ground parallel to the topography. They are more or less linear in plan, with some meanders, of course, but they hardly ever double back or form loops. Limestone caves develop along bedding planes and joints, and often show three-dimensional or maze patterns. For example, the straight-line distance between the ends of Ape Cave, one of the largest lava tubes in Washington, is about 9,100 feet. Its single, moderately sinuous passage would, if straightened out, measure about 11,200 feet. Oregon Cave, a limestone cave of near-maze configuration and the main attraction of Oregon Caves National Monument, occupies a space that measures only about 950 feet by 500 feet with a depth of 400 feet. But its passages, if straightened out, would total nearly 15,000 feet.

Typically lava caves are single, meandering, shallow passages five to fifty feet below the surface with little vertical relief. Lava cave entrances are almost always at points of roof collapse, while entrances to limestone caves take many forms. Because the walls and floors are extensively cracked and porous, lakes and streams are rare in lava caves. But the lava is rough! Surfaces are very abrasive, and boots and clothing wear out quickly.

Decoration is sparse in lava caves compared to those in limestone, but what speleothems there are stand out well against the dark walls and ceilings. Finally, limestone caves are often modified greatly by groundwater, but lava caves are essentially static once the flow that formed them ceases.

In the remote rangeland of Malheur County in southeastern Oregon lies a group of most unusual lava caves. They formed from a late Pleistocene lava flow which spread broadly to the east from a vent at the foot of the Sheepshead Mountains, about seventy miles southeast of Burns, Oregon. Evidence indicates that settlers and herdsmen have used the caves periodically for shelter and storage for a hundred years or more, and because of their attractiveness and closeness to Pleistocene lakes of the northern Great Basin it is probable that they were used by Indians for many hundreds of years previously. In addition, the lava tubes are populated by a large variety and number of animals and birds. Only recently have these caves come to the attention of recreational cavers and speleologists but already their value as a unique biological, geological, speleological, and cultural study area has been established.

The Saddle Butte system lies in a high, broad river basin known as the Owyhee Upland. The low relief gives little indication that below the plain lies one of the world's largest lava cave systems. The annual rainfall, about nine inches, supports some

The collapse of the ceiling formed this lava trench in the otherwise flat landscape of the Saddle Butte area. *Charlie and Jo Larson.*

grass and a lot of sagebrush. The country is mostly unfenced open rangeland grazed by beef cattle, herds of domesticated and wild horses, deer, and antelope. During World War II it was unofficially used as a firing range, and it is common to find 50-caliber brass casings and occasional live rounds.

Despite the lonely appearance wildlife abounds. Deer and antelope graze contentedly near cows and horses. Jackrabbits proliferate, in turn supporting a large coyote and bobcat population. Colonies of mice and packrats in cave entrances and overhangs attract many large birds, including crows, ravens, great-horned owls, and even eagles. Rattlesnakes are not uncommon, especially at cave entrances where they move in and out to achieve favorable temperatures. Bull snakes and gopher snakes have been observed far inside some of the caves. Remains of many other animals have been found in the caves, including weasels, badgers, marmots, burrowing animals, and the very rare desert kit fox, *Vulpes macrotus*, which is on the Endangered Species list, and has been reported only a dozen times in Oregon. Bats, singly and in small colonies, are commonly observed in all of the system's caves.

Every cave, particularly around the entrance, is an animal den. Hoofed animals find shelter in the sinks and trenches, but they seldom venture far into the caves unless the slopes are relatively smooth. Remains of large vertebrates, including horses, have been found far inside, but it is likely that they were dragged there by predators or scavengers. Predatory animals use all parts of the caves they can reach, but they are unable to negotiate vertical drops. Remains of large animals found beyond most vertical drops most likely represent a one-way trip. Small rodents stay close to the entrance.

The Saddle Butte system is the partially collapsed, central drainage channel of a late Pleistocene lava flow about five miles wide and nineteen miles long. The lava flowed to the east from a vent elevation of about 5,100 feet down to about 3,800 feet at the eastern edge. The known caves are found in the downslope eight miles of flow, which has a slope of about one degree—correlating well with similar lava caves elsewhere. The slope is fairly constant throughout the system except where it crosses a north-south trending fault near the eastern end. The fault, about one hundred feet high, hasn't been positively identified as either a pre- or post-flow feature.

Naming lava caves can get complicated, since the caves so often occur in groups. Every lava cave is a segment of a lava drainage system. Entrances are almost always at points of roof collapse, which may or may not isolate that segment from the rest of the system. Such collapses, if local and of moderate dimensions, are referred to as sinks; more extensive ones are called collapse trenches. Often sinks result in two entrances—to an upslope cave and a downslope cave—both of which can be

The largest cross-section of the system—fifty feet high, and ninety-eight feet wide—is in Owyhee River Cave. *Charlie and Jo Larson.*

considered the same cave. However, if such entrances are at the ends of a relatively long collapse trench, and especially if discovered at widely separated times, the respective caves receive separate names. Finally, a lava cave "system" is composed of these remaining, separately named, cavernous segments of a common lava drainage channel.

There are sixteen named caves in the Saddle Butte system: Baker, Bullet, Burns, Can, Coyote Trap, Fire Pit, Kitty Pooh II, Kitty Pooh Extension, One Station, Owyhee River, Patton's Arch, Raven Pit, Rattlesnake Number Two, Skylight Number Two, Tire Tube, and Tooth caves. Many are named for objects found in them. Total length of these is 11,900 feet. The longest cave of the series is Baker Cave with 4,800 feet of passage. The largest cross section is found in Owyhee River Cave east of the fault, where it is fifty feet high and ninety-eight feet wide. West

Jo Larson gazes at the strange shapes of the weathered basalt in Raven Pit Cave. *Charlie Larson.*

of the fault cross sections typically are as high as they are wide; east of the fault they tend to be considerably wider than high. The caves are developed on one level except for Baker, which has about 300 feet of an intact upper level west of the fault near its entrance.

The lower ends of Baker and Tire Tube are flooded by infrequent torrential rains about every three years which quickly drain away. Standing water has never been reported in any of the caves, although the mud floors in areas which flood can be quite moist.

Almost all of the original cave floors are obscured by breakdown, detritus, or—most commonly—dry and unconsolidated clay. The original cave walls are typically smooth basalt rock, but often they are modified by breakdown, erosion, or secondary mineralization. There are few original lava speleothems, but some impressive secondary mineralization from weathering and leaching can be seen. Raven Pit Cave, named for a raven's nest complete with young birds, is the most outstanding example of the strange weathering processes found in the system's caves. Weathering has also exposed some roof section jointing. Outside, the surface of the lava flow weathers slowly, as might be expected in a semi-arid climate. In the caves, however, evaporation of what little water that enters induces rapid decomposition of the basalt. Water combines with the minerals in the basalt to form secondary minerals. These usually occupy more space than the original constituents so that bits and flakes of basalt are literally pushed off the walls and ceilings. Since this decomposition is selective—concentrated around points of water entry—it produces grotesque shapes and contours.

Cave entrances are at points of collapse, and one can usually climb down over moderately inclined breakdown slopes to the silty clay or breakdown-covered floors of the caves themselves. Often the entrances are at the end of a collapsed trench. Entrance slopes show the effects of centuries of animal traffic, and the tops of rocks in the trail are polished glassy smooth by traffic. About half the caves end at breakdown obstructions; one ends at a lava plug; and the others end as flat clay floors meet descending roofs.

The caves of the Saddle Butte system differ markedly from

lava tubes elsewhere. For one thing, they are very dry, with a relative humidity as low as fifty percent inside the caves, exceptionally low for any kind of cave. Animal carcasses dry out before they decompose and the large quantities of animal droppings are seldom offensive, for they are quickly desiccated with no odor. In some caves foot traffic raises so much dry dust that taking photographs is very difficult. The caves are warm, which isn't remarkable, but the combination of warmth and low humidity makes them comfortable year round. On the surface it is often hot in the summer and cold and windy in the winter. Many cavers on weekend trips prefer to sleep in the caves rather than to set up tents on the surface.

During rare times of high humidity, the odor from the droppings of bobcats can be quite strong. My first visit to Owyhee River Cave was during one of these periods. We felt we were in an animal den fifty feet wide that went on for 1,200 feet because of the numerous animal beds surrounded by gnawed bones and bits of fur, the large quantity of animal droppings, and the oppressive musky smell. It gave us an eerie feeling that we were being watched by a thousand animal eyes, but the only life we actually saw in the cave on that trip was a lethargic, harmless bull snake.

No other lava caves are known to harbor either the variety or numbers of animals found in the system, most of which are not often known to utilize caves. That these caves are so biologically rich is due in part to their isolation, and that condition is likely to continue, for nowadays few people other than spelunkers and speleologists have any interest in them. The land is utilized only for open range, and there is no push for improvement of the miserable roads.

Modern cavers became interested in the area in the 1960s. In 1962, a businessman in the small settlement of Jordan Valley told a caver about several large caves near Rome, Oregon, which required a rope for entry. At that time there were few cavers around, and there were more promising leads to look into, so no speleological investigation took place. Several years later a Portland caver who was reviewing recent aerial photographs confirmed the existence of a chain of caves in the vicinity of Saddle Butte. This spurred interest, and the Oregon Grotto of the National Speleological Society scheduled a field trip to the area for Memorial Day, 1967. But some anticipated sources of information didn't develop, and the cavers had lots of other attractive cave leads. So the trip didn't occur.

In June 1967, the Oregon-Idaho caving community was startled to read a rapid-fire series of newspaper articles which described a "Forty-Mile-Long Cave" just discovered in southeast-

The lower level in Baker Cave begins at the foot of the ladder, extending back toward the camera. Multi-levels in lava tubes indicate two or more separate lava flows. *Charlie and Jo Larson.*

ern Oregon. The Bureau of Land Management, which apparently commissioned the aerial photographs, had discovered the caves. Stimulated by exciting but redundant headlines such as "Tales of Giant Underground Malheur County Cavern Verified by BLM Officials!" a rash of planned caving trips broke out.

The "discovery," spurred by the local newspapers, reached the wires of the Associated Press and United Press who dutifully repeated the forty-mile-long exaggeration. That figure began with local citizens, who recalled that some old-timers had surmised there was a chain of caves which might be that long. Another well-known lava cave, Malheur Cave, lies about thirty-five highway miles northwest of Saddle Butte, and perhaps in earlier days when the true origin and extent of the caves was much less understood, that cave and the Saddle Butte caves were thought to be the same. A month later newspaper accounts of the find had shrunk to a more reasonable length of eight-and-one-half miles of partially collapsed lava tube—with the longest cave pegged at 4,000 feet—and they acknowledged the rediscovery of a cave system known locally for more than a century. Cavers everywhere have sadly learned the truth of the axiom: "Divide all cave rumors by one hundred"—but in this case it was only a five-fold exaggeration.

Far from being "new" when rediscovered by the BLM, the caves were known to locals and transients since at least the late 1800s. Early stockmen used the caves as stopover points during cattle drives across Oregon for their relatively dry, warm shelter, and the storage they provided for provisions and grain. Several sturdy wooden boxes from a St. Louis, Missouri, provisioning company were found in Baker Cave.

Cattle drives usually just passed through the area, but sheepherders and their flocks stayed on. Semi-permanent camps were set up which utilized the open lava trenches, sometimes augmented with stone fences as corrals. Adjacent caves became shelters for the sheepherders. Sheep raising in the Owyhee region is carried out mostly by Basques, a small, closely knit ethnic group from the Pyrenees Mountains of Spain who came to the area in the late 1800s. For them, the many miles of lava trench and caves provided a ready-made chain of shelters and corrals on the way to and from remote grazing areas. As an aid to navigation through the locally featureless rangeland—and no doubt because the sheepherders in this lonely place had time on their hands—they carefully constructed large stone cairns. Each cairn, now called a "stoneman," has a distinctive appearance. They were strategically placed, singly and in groups, to identify specific locations or routes. For example, three stonemen close together mark where the lava tube chain crosses the first road through the area. Others, spaced at wide intervals, mark apparent throughways, not necessarily cave entrances, and they are skilfully aligned in a slight arc so that intervening stonemen don't obscure more distant ones.

The stonemen weren't the only improvements made by these cave-dwelling sheepherders. Cairns even larger than those

Basque sheepherders used many of the cave entrances for shelter and built stone fences for corrals. A cairn, or stoneman, in the background marks this entrance. *Charlie and Jo Larson.*

on the surface were built to provide easy access to caves with vertical entrances. Stone fireplaces and windbreaks are found here and there. Many shelters still have pothooks and wires in place over firepits. Rusty provision cans, five-gallon rectangular water tins, and even an occasional old newspaper, are half-buried in the sandy floors.

Hunters and trappers soon realized that the caves were inhabited by a variety of animals, some valuable for their pelts and others for the bounty offered by the state. Clarence Eckstein, for years owner-manager of the tiny nearest settlement, trapped cats and coyotes in the system's caves. He said it paid well: in 1926 coyote pelts brought up to twenty-five dollars each, while the lumber camps farther west paid five dollars for a day's work. Today hunters and trappers are still drawn to the caves in search of bobcats, both for pelt and bounty, and increasing numbers of cats have perished inside the caves where they apparently fled

Still legible after ninety-two years underground when discovered in 1975, this register recorded visits to Burns Cave in 1883 and 1903. *Charlie and Jo Larson.*

after being wounded or poisoned. Some, too young to fend for themselves, starved when their mothers failed to return.

The most disgusting of man's use of cave systems was as a repository for rubbish. One sink became a garbage dump simply because it was near the system's first road. Now the nearest road still in use is several rocky miles away, and dumping has ceased.

It is likely that the cave system was used by ancient Indians for shelter and storage. One lobe of a large Pleistocene lake, around which Indians are known to have lived, was less than twenty-five miles to the west. Indians lived in caves until recently along the Owyhee River, which lies less than three miles from the lower end of the cave system. Add to this the relative attractiveness of these caves, and a strong argument for ancient habitation can be made. There have been no supervised archaeological excavations, but amateurs have dug holes here and there in the system's caves.

The earliest known cave register in Oregon, with two entries, was found at the rear of Burns Cave. A folded note was wedged among the rocks atop a cairn. The unprotected paper was well preserved in 1975 after ninety-two years in the cave, a tribute to the low humidity. Despite attempts by rodents to devour the paper, giving it a ragged appearance when unfolded, the following was still legible:

> April 10th '83
> Explored by Spencer,
> Kelsoe and Shehan
> ?. E. Smith
> W. Hooper
> Jan 3, 1903

Other turn-of-the-century visitors left notes in a mayonnaise jar at the back of Owyhee River Cave. Earliest of these was "Arego Harrison, February 6, 1897." Harrison was known to be a camp tender for a sheep outfit. Early explorers used torches constructed of twisted wire with a rag soaked in turpentine or kerosene at the end. Until a few years ago several of these torches were still seen at various entrances. In Baker Cave, a pint whiskey bottle half-full of turpentine was secreted behind a rock. The bottle bore a still-legible tax stamp dated 1911.

Beyond the fascination of the caves themselves, the Saddle Butte area is a great place to go caving. It's a quiet place. The daytime silence is broken only by the occasional passage of a jet overhead, or the call of a crow; at night only the coyote's howl breaks the peace. It seldom rains. Summer days are hot but bearable because of the low humidity. Nevertheless, no matter how tolerable the daytime heat one soon learns to anticipate the cool nights. In winter, when it is cold and windy, the caves are a welcome relief—warm and dry. And, best of all, the abominable roads are likely to keep the area the way it is.

Floyd Collins, Hero of Sand Cave

Floyd Collins spent a lifetime exploring caves in the Mammoth Cave region of Kentucky. Although he was the foremost cave explorer of his time, he was relatively unknown until the last two weeks of his life in 1925. Those two weeks Collins spent trapped underground, his foot caught by a rock, slowly dying despite a massive rescue effort. Overnight Collins attracted nationwide front-page headlines, and this public interest was sustained even beyond his death. What factors focused and maintained the public interest in what was in reality a relatively minor incident?

Kay F. Reinartz became interested in Floyd Collins after several visits to Floyd Collins' Crystal Cave. She has published a variety of articles on spelean history and other subjects.

Floyd Collins, Hero of Sand Cave

KAY F. REINARTZ

Over fifty years ago an unknown Kentucky farmer named Floyd Collins became a national hero overnight when he was trapped in a small cave near Mammoth Cave, Kentucky. Collins, thirty-five, a highly experienced underground explorer whose family ran nearby Crystal Cave, was working for a neighbor who hoped that a cavern with tourist value might be found on his land. In fact, when Collins entered Sand Cave on January 30, 1925, he had a contract guaranteeing him fifty percent of all profits realized from his explorations. But within a few hundred feet of the entrance a small boulder dislodged from the ceiling, pinning his foot. After almost a week of futile effort to get Collins out through the narrow natural passage, a cave-in cut him off from rescue workers. A shaft from the surface was begun and for eleven days volunteers dug in cold rain and mud until they finally reached the man, only to find him dead.

For two anxious weeks Collins was the most written-about, worried-over, talked-about, and prayed-for individual in the

---

Floyd's coffin in the Grand Hall of Floyd Collins' Crystal Cave. *Thomas C. Barr, Jr.*

United States. His story quickly captivated the public and has continued to fascinate in the years since his death. In constrast to most popular heroes, who are lionized for a glorious voluntary feat, Floyd Collins became a hero not for how he lived but for how he was forced to die.

Others have been trapped in holes in the ground during prolonged rescue operations, but none has received the attention accorded Collins. Just two months after his accident seventy-three miners were caught underground at Coal Glen, North Carolina. Although fifty-two men died, the story was given only minor coverage after the first day. In April 1949, national attention was focused on a three-year-old California girl, Kathy Fiscus, during unsuccessful efforts to rescue her from a fourteen-foot well. While this three-day incident gained more attention than most similar cases, it failed to sustain public interest, and the memory of the tragedy faded with the retrieved body. In these and other instances the memory of Floyd Collins has been frequently invoked by the press. It is Collins' entrapment that stands out as the classic case.

The Man Buried Alive

The most important feature of the Collins story responsible for capturing and holding public interest and sympathy was undoubtedly the man himself: from the very beginning Floyd Collins was a vibrant human being buried alive. This perception of Collins was developed by William Burke Miller, a twenty-one-year-old cub reporter for the Louisville *Courier-Journal*, who told Collins' story to the public with great human sensitivity. Miller was dispatched to the cave Monday morning, February 2, 1925, and promptly became directly involved in the rescue. A small man able to navigate the constricted hundred-and-fifty-foot passageway to where Collins lay, Miller first entered the cave out of curiosity. After meeting Floyd Collins face to face and seeing his desperate predicament, he developed great respect and liking for the man. In the following days Miller worked tirelessly for Collins' release, taking him food and drink, and leading rescue teams.

Miller reported his conversations with Collins, as well as his own reactions to the cave environment, in the press. After his initial trip into the cave, Miller told the reading public: "I felt I never wanted to go back in. The darkness, the damp, the helplessness of Floyd Collins, all filled me with horror." On February 4, he described the cave as "terrible inside, the cold, dirty, water numbs us as soon as we start in. We have come to dread it, but each of us tell ourselves that our suffering is as nothing compared to Collins."

The physical state of the exhausted rescue workers gave mute testimony to the conditions Collins was enduring. Every man emerged from the cave bruised, cut, and water-soaked. In the first five days six cases of pneumonia and three of diphtheria developed among the hundred-odd workers. During the long attempted rescue the newspapers recorded the physical and mental collapse of scores of workers, including Floyd Collins' brothers, professional miners, and Miller himself.

Miller's reporting of Floyd Collins' own words helped the public to identify with him as a living man helplessly trapped and totally alone in a dark, wet, cold, underground prison. While Collins was brave, he was portrayed not merely as a paragon of courage, but as a vulnerable man filled with ambivalent feelings about the desperation of his plight. On February 2 he told Miller:

> Four days down here and no nearer freedom than I was the first day. How will it end? Will I get out or—? I couldn't think of it. I have faced death before. It doesn't frighten me. But it is so long. Oh, God be merciful. . . . I know I am going to get out. I feel like it. Something tells me to be brave and I am going to be.

Through Miller, readers saw Collins as a warm, sympathetic person capable of showing concern for his rescuers in spite of his own pain and agony. On February 3, after Miller had been digging out rocks and dirt from around Collins' body with his bare hands for hours, Collins told him:

> Now fellow (this is what he calls me) you better go out and get warm. But come back. . . . I want you to tell everybody

> outside that I love everyone of them and I'm happy because so many are trying to help me. Tell them I am not going to give up, that I am going to fight and be patient and never forget them.

Undoubtedly Collins' frequent acknowledgment of public concern and his expressions of gratitude drew Americans to the man.

Collins' physical and psychological stamina throughout the ordeal was phenomenal. On February 4 Miller told of the impact of the trapped man's personal strength. "His patience during long hours of agony, his constant hope when life seemed nearing an end, is enough to strengthen the heart of anyone. Collins doesn't know it, but he is playing a very, very big part in his own rescue." Collins' dogged optimism and personal courage seemed to infect all of the workers. In spite of the increasing hopelessness of the situation, a "we-won't-give-up!" atmosphere predominated at Sand Cave.

Although Miller may have stretched the journalist's rule requiring absolute objectivity in reporting the news, he created an honest, compassionate story of a moving human experience. For his stories on Floyd Collins he was awarded the Pulitzer Prize in 1926 "for the best example of a reporter's work during the year, the test being strict accuracy, terseness, and the accomplishment of some public good commanding public attention and respect."

The Media and the Big Story

To state that Floyd Collins' acclaim as a hero was a product of the times or Miller's outstanding reporting is to oversimplify the issue. As popular culturist Marshall W. Fishwick has observed, "Behind every hero is a group of skillful and faithful manipulators." In Collins' case the creators of the myth were the media, particularly the press, demonstrating the power to excite millions over a relatively small matter. Modern technology, permitting rapid transmittal of news, also contributed greatly to the development and maintenance of a high level of public interest during the prolonged rescue operations.

In forty-eight hours the story of Floyd Collins' accident was transformed from a Kentucky human-interest item to *the* big story with page-one streamer headlines nationwide. For two weeks large newspapers such as the Chicago *Tribune* printed average daily coverage of three full columns, complete with diagrams of Sand Cave, rescue plans, and dozens of photographs. Newspapers sold out as fast as they came off the presses. One New York City newspaper expanded daily sales by more than 100,000 papers and other papers in the city increased circulation by 50,000 to 60,000. Newspaper sales records for this century show that, in the era between the World Wars, only Charles Lindbergh stories had a greater impact.

Masses of printed material seemed only to whet the appetite of a totally involved public. People wanting more and later news kept newspaper office phones ringing around the clock. The Chicago *Tribune* chief operator estimated that from the very start they received at least 4,000 telephone inquiries about Collins daily.

In 1925 radio was new enough to be marveled over, but well enough established to be an effective means of disseminating news to the public. Although news reports were an important part of radio programming, special bulletins between regular newscasts were unprecedented. This was changed on February 11, the thirteenth day of the rescue, when Station WHAS, Louisville, began issuing spot bulletins on developments at Sand Cave. Other stations followed suit. Widespread radio coverage helped increase newspaper sales, as people who heard the story over the air bought papers to learn more.

Across the country the latest bulletins on Floyd Collins' rescue were publicly displayed. Movie theaters, hotels, and restaurants posted a "Collins Report" for their guests. Some theater performances were interrupted to announce new developments at Sand Cave.

Even though Floyd Collins' story carried a great deal of psychological and emotional appeal for the public, its success in capturing the headlines was helped because it came at a slow time. Others have also observed this, including Roger Brucker in "The Death of Floyd Collins" in *Celebrated American Caves*. Immediately prior to Collins' accident headline stories included the arrival in New York of the German dirigible *Los Angeles*, the resignation of Charles Evans Hughes as secretary of state, Gloria Swanson's third marriage, and the discovery that President Calvin Coolidge exercised daily on a hobby horse in the White House.

The story of Floyd Collins unfolded day by day from the more than two hundred reporters on the scene. Initially reporters hung around Sand Cave just long enough to secure a story. They then raced into Cave City, often to stand in line for hours waiting to wire the story on the few available lines to their home offices. This congestion was somewhat relieved when a Western Union station was established outside the cave with twenty operators on duty around the clock. Rapid news transmittal was further facilitated by the installation of dozens of telephones on trees near the cave and by the arrival of two fully equipped "ham" radio operators from Indiana.

Although the press managed to fill column after column, there were only a few facts involved in the entire affair; most of

Floyd Collins as he appeared a few months before his death. *Courier-Journal and Louisville Times.*

the miles of words printed were elaborations of these facts plus sidelight "human-interest" stories. Few reporters had the first-hand story as did Miller, and even fewer exercised his restraint and sensitivity. Reporters rewrote Miller's stories with much embellishment, and shamelessly pirated material from each other. Often dispatches from Sand Cave were an indiscriminate mixture of facts and purple prose designed more to excite than to inform. Competition among reporters was intense. They turned to anecdotal material on Collins' life and family, general local color, and cave-site anecdotes for more story material. The stories often contradicted one another, and it was difficult for the reader to determine what really was happening. An anxious public was alternately informed in screaming headlines, "ALL HOPE FOR COLLINS RESCUE ALIVE GONE," and then "COLLINS OUT BY MORNING." This helped maintain a high level of tension in the case—exactly what the press wanted—but for the average person it was keenly distressing.

The Sand Cave Carnival

Soon after rescue work began at Sand Cave crowds of gawkers gathered. Bootleg liquor appeared and loud arguments and fistfights broke out over the best way to get Collins out. Local residents in control of the operation were quite hostile to suggestions and assistance from outsiders, and when technical experts arrived offering their aid, such as a team of Louisville stonecutters, they were turned away. Disorganized rescue crews went in and out of the cave day and night, but little progress was made.

Collins was communicating daily with his rescuers, and his greatest threat at this time was deterioration from hypothermia—serious loss of body heat. The main means of fighting death under such circumstances is to keep the body of the victim warm and to maintain physical and psychological strength through warm, nourishing food and drink. When Lieutenant Robert Burdon of the Louisville Fire Department crawled down to Collins he noticed numerous bottles of coffee and parcels of food scattered along the floor or stuffed into crevices. Miller observed this too, as well as broken glass in the passage. Much of the food intended for Collins obviously never reached him. Many volunteered to take nourishment to the trapped man. Frightened once they entered the cave, these would-be heroes hid the food, emerging later to announce to all that they had indeed fed Collins. Meanwhile, far below, Collins was slowly deteriorating from hunger and cold.

Lack of definite leadership contributed to the general inadequacy of the rescue. This was resolved when Henry Fields, Governor of Kentucky, placed Lieutenant-Governor H. H. Denhardt in charge. With the help of fifty heavily armed national guardsmen Denhardt quickly restored order. His first act was to bar all local people from participating in the rescue work, threatening to arrest those who did not comply. He placed the operation completely in the hands of mining engineers and the Red Cross. Quantities of heavy equipment and supplies were brought in, including 2,400 rounds of ammunition. The Governor of Tennessee expressed his concern by sending a handful of soldiers to aid the Kentuckians.

The guards were needed to help maintain order in town as well as at the cave. The population of Cave City quadrupled to almost 3,000 people from its normal 690, with hundreds more pouring in daily. More than 2,000 persons visited the rescue camp daily. One reporter noted that cars came and went at a rate of 120 per hour, and he counted license plates from more than twenty states.

The hills around Sand Cave were a study in disorder. Trucks bumped over the rough road from Cave City haphazardly dumping miscellaneous machinery and supplies. Sledgehammers, logs, picks, wire cables, coffeepots, and lumber were thrown about helter skelter. Farmers, engineers, miners, college students, plain muckers, and hillbillies crowded around the black hole to send off and greet each rescue crew.

On Sunday, February 8, after a week of front-page coverage, a crowd variously estimated at from 10,000 to 50,000 made its way through mud and water to Sand Cave to find it surrounded by barbed wire and armed soldiers. However, a full-scale tent city spread out before them. Clustered around the cave entrance were a power plant, powder magazine, blacksmith shop, a fully staffed field hospital, barracks, and rest hall. For visitors there were lunch and fruit stands, a restaurant, and a taxi stand. As thousands of noisy tourists wandered aimlessly through the tent city, a long line of silent, hollow-eyed, grimy men filed into the Red Cross canteen for a free hot meal.

By midday the entire scene resembled a carnival rather than the site of a life-and-death struggle. Enterprising local residents as well as outside profiteers provided the milling crowds with hot dogs, apples, soda pop, and sandwiches, at five times the standard 1925 price—twenty-five cents for a tiny hamburger. An old-fashioned medicine man in a covered wagon exhorted one and all to "come and be cured." Little knots of excitement were generated around small-stake gambling games. Jugglers and sleight-of-hand artists passed through the crowd, offering diversion to people bored and frustrated with their inability to observe actual rescue operations.

Fundamentalist preachers mounted stumps to rant and pray for Floyd's deliverance, while con artists took up collections "to buy food and supplies for the volunteer workers trying to rescue Floyd." At 3:30 P.M. holy diversion was provided by Parson James A. Hamilton of Louisville, who performed "officially approved union services." The papers described Parson Hamilton as "just a long, lean circuit rider," who rode twenty miles on a sleepy mule to bring "the word of hope to people of the same simple faith." Across the country congregations joined in special Floyd Collins services, often dedicating the collection to the rescue fund. From the White House, President Coolidge expressed his concern and hope, and sent prayers. Many remained after dark, silent and shivering under the glare of light bulbs strung from tree to tent pole.

Local Color and the Collins Family

The press exploited, often to the point of caricature, the Kentucky hill country social milieu and the Collins family's folk image. They were depicted as simple, fundamentalist, hardworking backwoods people who eked out a marginal existence on the family farm. Floyd's father, Lee, sixty-two, was the most popular family member—probably because he was the most willing to talk with reporters. The press reported in detail the old man's antics at Sand Cave, his philosophy, and his comments on Floyd's love of caving. The image created was that of a comic-pathetic old man with a tobacco-stained grey stubble beard dressed in an old red sweater and worn army coat who moved through the crowds reciting Scripture to anyone who would listen. "Blessed be the name of the Lord: the Lord giveth an' the Lord taketh away," he would quote to his listener—and then give him a handbill advertising nearby family-run Crystal Cave. The admission was a steep $2—a sizeable increase from the normal fee—but because Crystal was Floyd's cave and people were bored and looking for distractions, many paid Lee Collins' price and made the trip. By the end of the rescue Lee Collins had changed the name of the family cavern to Floyd Collins' Crystal Cave.

Floyd's stepmother, Jane Collins, had a premonition of his entombment, which duly appeared in the San Francisco *Chronicle* February 6, allegedly as she told it:

> An' Floyd said to me "Lor', Ma, I got three days more work in that cave, and Lor' how I wish it was over. I been a-dreamin' of bein' caught in some rocks and some men a-clawin' at me." And I sez to him: "You stay home here today. We ain't got no wood chopped, and we need you here." Well suh, that boy he went down behind the house and chopped up a whole pile of wood. But at 10 o'clock he went away. That night he didn't come back.

Other papers added that in the dream Floyd found that he was suddenly surrounded by angels. Although Floyd Collins was not known to be a religious man until his fatal accident, angels seemed to hold a special fascination for him. He reported to Miller during their first meeting that he dreamed that angels on white horses brought him white chicken sandwiches.

Public Response

A flood of calls, letters, and telegrams offering spiritual and material help poured in. Many were addressed directly to Floyd

Collins. One telegram from a man in New York City was delivered sixty-five feet underground. "I am praying for you, old timer," the sender wrote. "I'm betting your grit will pull you through." The St. Louis *Post-Dispatch* claimed that Collins asked his brothers to give the sender a gallon of milk and some stewed onions—his favorite dish. Well-wishers repeatedly advised him "be courageous, calm, and don't worry," and "trust in God, and all will be well. You will be delivered." Miller informed the public of Collins' response: "It's mighty nice to know so many people are pulling for me. Tell them I love them all. I am not afraid to die. I believe in heaven. But I believe I *am* going to be taken out alive."

The ambiguity of Floyd Collins' situation was a source of agony. Workers at the site could freely communicate with the trapped man, but could not get him out. Many were convinced all that was needed was the right technique, and they hastened to send advice and equipment. A Kansas citizen advised "use a small electric hand drill," and an Elmira, New York, man recommended the use of two railroad jacks. A Blacksburg, South Carolina, man wired Homer Collins: "I have a man who can free your brother with a strapper machine. Get strapper and wire if you want this man to come." Chicago millionaire-philanthropist Mrs. Emmons McCormick Blaine, remembered Floyd as her cave guide the previous summer and sent, via private plane, two doctors fully equipped to amputate his leg. A tourist party from Houston that Floyd had guided offered a $1,000 reward to anyone with a successful plan for getting Collins out of the cave. While these offers provided their senders some psychological relief, most of the thousands of suggestions were valueless. Undoubtedly the most useful aid was that sent to feed and equip the hundreds of volunteer workers. Thousands of dollars in cash and goods were sent to the camp during the two weeks of the operation.

Meanwhile, Back at the Cave...

While the nation wrung its hands over Floyd Collins, the rescue work at Sand Cave continued as a shaft was sunk under the direction of Henry T. Carmichael, a local tunnel expert. Diggers were divided into teams of four men, each working a half hour at a time in the pit. Volunteers of every age and background came to help dig. Strong-backed Nashville railroad hands worked alongside students from Vanderbilt University. The Bowling Green Teachers College football team came and immediately enterd into a good-natured contest with a gang of Louisville boys over who could move the most rock and mud in the allotted work time. A mine-disaster team arrived from the U.S. Bureau of Mines. A crude hand winch was installed to hoist debris from the deepening shaft as the crews energetically worked to harmonica music or the chant, "Dig! Dump! Pray!" In spite of the desperation of the task and the cold mud of the shaft, morale was generally high.

One of the volunteer workers, sixteen-year-old Eddie Bray, was in Hot Springs, Arkansas, shining shoes for his breakfast when he heard about Floyd Collins. Hitchhiking to Sand Cave, he worked for eleven days on a shaft crew. He was reported in the press to be "Spark Plug Bray," a professional welterweight fighter, who allegedly had been training in Hot Springs before coming to Sand Cave. Actually, the full extent of Bray's fighting experience was sparring in the ring on a single occasion. After the rescue was over Bray toured movie houses in Kentucky and Indiana relating his experiences in the tragedy.

As the shaft digging dragged on for eleven monotonous days, reporters searched frantically for fresh material. It was announced by Miller that Collins might have been electrocuted by a high-power line used for shaft lighting when it came into contact with the wet wires leading to the bulb hung around his neck for light and warmth. Later the public was assured that radio circuit tests based on electrical impulses from the same bulb indicated that he was still alive. On February 11 banana-oil fumes were blown into Sand Cave as hundreds of sniffers scattered over the landscape seeking a hint of odor that might indicate the presence of an alternate entrance. Stories were dedicated to the famous Kentucky bloodhound, Joe Wheeler, hero of a hundred manhunts, who was sent unsuccessfully down the natural cave passage with a canteen of water hanging from his neck in the hope that he could squeeze past the cave-in to Collins. A tragic sidelight story involved a thirteen-year-old boy in Barnesboro, Pennsylvania, who was crushed to death while playing "Collins in the cave" in an abandoned mine.

Publicity seekers across the country sought fame by associating themselves with the hero Floyd Collins. A man in Haddam, Kansas, sent Miller the following telegram on February 4:

> Please contradict statements that I am buried alive in Sand Cave. Tell mother I am all right. Am coming home.
> Signed, Floyd Collins

Haddam Mayor F. W. Shearborn wired a physical description of the penniless man to Lieutenant-Governor Denhardt—collect. Tattoos quickly established that the man was an impostor. Although thrown into the Haddam jail on vagrancy charges, he managed to wrangle a newspaper reporter out of $50 for an

Rescue workers toiled around the clock to remove debris from Sand Cave. *Courier-Journal and Louisville Times.*

exclusive story by signing an affidavit verifying that he was the original Floyd Collins.

Few stories have universal appeal without some romance, and the press managed to develop this angle. Thirty-five-year-old bachelor Floyd Collins was not known to have a sweetheart, but at least three young women in the area claimed they were engaged to him. Newspapers told about Collins' "little girl," twenty-two-year-old Alma Clark from nearby Horse Cave, Kentucky, with pictures of the pretty, calico-dressed young woman. Miss Clark maintained that she and Floyd were to have eloped on February 5. Women elsewhere rushed to link themselves with the buried man, and many letters contained offers of marriage. One adventurous miss proposed that she and a minister go into the cave and sanctify the marriage at once, in spite of Collins' diminishing chances of survival.

From the beginning rumors had circulated that the entire story was a publicity stunt. Reputedly Collins either left the cave each night by a secret passage, or he never was in the cave at all. An official Court of Inquiry was established by Governor Fields in Cave City to investigate these rumors, as well as claims that deliberate efforts had been made by individuals and groups to interfere with the rescue work. The Court of Inquiry placed the blame for spreading false rumors on Tom Killian of the Chicago *Tribune*, who was summarily recalled by his home office. The Associated Press was vindicated by Lieutenant-Governor Denhardt, and the press was generally praised for honest, accurate reporting. The Court closed and the mobs surged back to the cave site.

The scene at Sand Cave on the second Sunday of the rescue was described by one reporter as resembling "a great political convention or a small war." By noon the Cave City road was immovably blocked, as owners abandoned their cars and proceeded to the cave area on foot. Estimates of the crowd were as high as 10,000 cars and 50,000 people. The carnival atmosphere of the previous Sunday again prevailed. Excitement was high, as the shaft was expected to reach Collins at any minute.

The breakthrough came the next day, February 16. The news that millions had waited seventeen agonizing days to hear was flashed around the world: "DEAD!" Everywhere it brought mixed feelings of sadness and relief. Because of the crumbling condition of the shaft, the body was left where it was. A coroner's jury of six slid one by one down the shaft and through a short tunnel to identify the body.

On Tuesday, February 17 open-air funeral services were held on a bluff above Sand Cave. Five local ministers participated in the service, attended by 150 persons who sat on rocks and logs. The Reverend Roy Biser climbed on a stump to share these last thoughts:

> Floyd loved the caverns and caves, loved them as some of us love the flowers and the birds. Now he is enshrined in his sarcophagus of stone, where his body lies in peace. No other incident within memory has brought so many prayers from the brotherhood of man for one fellow man.

A local undertaker dropped a piece of ash, a small fern, and a bit of earth down the cave. As Floyd Collins' unseen remains were committed to his Maker movie cameras whirred and shutters clicked. Reverend C. K. Dickey of Horse Cave noted that "heroic deeds have laid a permanent monument in the exhibitions of courage and stamina revealed for seventeen days at Sand Cave." Unshaven grimy rescue workers stood with Floyd's neighbors and friends, bareheaded in the penetrating winter wind and sang "Nearer My God to Thee," among dozens of unattended tree-station telephones.

By sundown the next day the shaft and natural passage to Sand Cave were sealed with concrete and rock. Lee Collins, looking down at the final closing of his son's tomb observed, "It is a fitting place, just like a church, for every cave was a church to Floyd. If he were here to tell us, he would say, 'I am content.'" A wreath sent by Louisville newsboys was placed at the entrance to Sand Cave, and everyone went home.

The Posthumous Travels of Floyd Collins

While the rescue was over, the legend of Floyd Collins of Sand Cave had only begun. Citizens of Cave City agreed with the Collins family that Floyd should be removed from the cave and given a proper burial, and they helped raise the needed $2,500. On April 22, 1925, the body was recovered. Restored by local undertakers, it lay in state for viewing for several days in Cave City. It was then buried in an ordinary grave on a hillside overlooking the now renamed Floyd Collins' Crystal Cave.

In early 1927 the Collins farm, including Crystal Cave, was sold to H. B. Thomas. He promptly exhumed Collins' body and put it on display in a glass-topped coffin in the Grand Hall of Crystal Cave. The Collins brothers naturally resented this, but a lawsuit to stop it was unsuccessful. In the spring of 1929 the body was stolen from its coffin and thrown over a nearby cliff into the Green River. Responsibility for this act was never established, but it was widely assumed that competitive cave owners did the deed, probably with the hope that tourists would be less interested in visiting Floyd Collins' Crystal Cave if there were no body to view, and thus more interested in visiting their own caves. This plan to dispose of Floyd's body was thwarted when it caught on trees below the cliff. The body was retrieved—minus one leg—and returned to its coffin in the cave, where it continues to rest today. The coffin can still be seen—closed, but with a tombstone—in the cave, now a part of Mammoth Cave National Park.

The Story Is Told Again, and Again, and Again

Before the Sand Cave tent city had been removed the story of Floyd Collins was being retold for profit. In the fifty years since, it has been the subject of films, lectures, songs, poems, books, articles, and television specials. At times the Collins story has merely provided the framework for a sellable product. In others, the story has been used as a vehicle for technical analysis, social commentary, or moralizing. In every instance it is the basic story, Collins-in-the-cave, that continues to interest people.

In the pre-television days of the 1920s the chief means of prolonging public involvement was through films and lectures. Just thirteen days after Collins was trapped, a picture produced by Cliff Roney, photographer for the Louisville Film Company, opened at the Alamo Theater in Louisville. The marquee advertised that it featured "The cave, the Collins family, the wild surroundings of the cave country, 'Skeets' Miller, the *Courier-Journal*'s hero-reporter, the Red Cross camp, the state directors of the rescue work, and many other scenes and incidents in the thrilling race against death." The announced purpose of the film was to show Americans "what Kentuckians are doing to save one of their own." In all, thousands of feet of film shot at Sand Cave were widely circulated in newsreels and specials. The impact of film in publicizing the incident was enormous.

Almost anyone involved in the rescue who cared to rent a hall could count on a crowd willing to pay double and triple the going rates to hear a "real, live account" of Floyd Collins at Sand Cave. A. B. Marshall, one of the few who actually reached Collins

Floyd Collins' tombstone and coffin. *Courier-Journal and Louisville Times*.

during the original effort, had removed artifacts from the body, including a small hammer, spade, and pocket knife. Rejecting fabulous offers for these objects, Marshall used them as the focus of his lectures. Harden T. Weaver, a Los Angeles miner who was one of the six men hired to remove Collins' body from the cave, toured movie houses throughout the Midwest and East. Before the feature film Weaver took over the stage to tell—as his handbills promised—how "Floyd Collins Could Easily Have Been Taken From Sand Cave, Ky., Alive and Unhurt." With the aid of moving pictures, stereopticon views, drawings, and Collins' words—as reported by Miller—Weaver let his audiences relive the agonizing weeks of the rescue. Weaver highlighted his talk with a number of objects, including the rope that had been tied around Collins' body in an early attempt to pull him out by force, and the first flashlight that illuminated his face. Weaver also displayed a light bulb which he claimed had hung around the trapped man's neck and a telephone he allegedly used to notify rescuers that he was still conscious. In reality, there never had been a telephone installed where Collins could reach it, and the original crew reaching the body in February found the light bulb broken. Weaver's enlightening lecture, together with the regular picture program, could be attended for only one dollar—four times the regular movie price.

The box-office potential brought a flood of offers to Floyd's family. Lee and Homer Collins signed up to recount the story of Floyd's life and death, and played to packed tents and halls. Hostility developed between father and son while each was independently making the rounds of the autumn fairs. When the Indiana State Fair opened in September 1925, Lee Collins was featured in the sideshow. Homer soon arrived and attempted to have his father's show closed. He claimed the old man was getting money under false pretenses and disgracing the family by soliciting contributions to pay off the mortgage—allegedly incurred during the effort to get Floyd out of Sand Cave. Homer declared that there was no mortgage. Lee Collins and his manager denied that they were using a mortgage appeal, and the show went on for two lucrative weeks. Homer angrily left town, returning to open his own film-talk at a local theater during the second week of the fair. He declared that he needed money to raise a monument to Floyd. The family feud was fully reported in detail on the front pages of the local newspapers as well as to the nation in *Variety*.

While a goodly profit was realized by vaudeville managers, the song industry really capitalized on the tragedy. In the spring of 1925 Polk C. Brockman, a record distributor and talent scout, keenly aware of the enormous record sales realized by tragedies, decided to market a song about Floyd Collins. He approached Andrew B. Jenkins, a blind itinerant evangelist-song writer from Atlanta. It is said that when Jenkins received Brockman's request he sat down at the piano and wrote the song in less than an hour. His stepdaughter Irene Spain arranged the music, basically a hymn melody, and Brockman paid them $25 for their effort. Brockman sold the song to Columbia Records who assigned it to Vernon Dalhart, a country-music artist. At least six other artists also recorded the song. It was an immediate hit, and for two years "The Death of Floyd Collins" outsold all other contemporary country-western records with sales in the millions.

A passage far inside Floyd Collins' Crystal Cave. *Thomas C. Barr, Jr.*

Dalhart's version appeared on nineteen labels, marking the beginning of his success as a country singer.

The ballad of Floyd Collins was quickly assimilated into folk culture and variations soon began to appear. Folk versions generally stress the rescue work and the moral to be learned from the tragedy. Barb MacLeod's 1972 ballad, "Grand Kentucky Junction," celebrates the long-sought connection between Mammoth and Floyd Collins' Crystal Cave.

Two books appeared soon after the Floyd Collins tragedy, E.D. Lee's *The Official Story of Sand Cave* and Howard W. Hartley's *The Tragedy of Sand Cave.* Hartley, a reporter for the Louisville *Post*, had his manuscript completed ten days after the sealing of the shaft. Dedicated to "Carmichael, the bronze hero of Sand Cavern," it provides a detailed account of the accident and the rescue work. In spite of biases—Hartley takes sides in controversial issues and lionizes Carmichael—this profusely illustrated work remains the best single source of information on the actual rescue operation and has been the basis for many later articles.

Short pieces have appeared occasionally in popular publications, ranging from books on caving, to the *Reader's Digest* and magazines such as *Saga* and *Man*. In many cases the story is grossly distorted and sensationalized. Notable exceptions are several pieces by William B. Miller. The fiftieth anniversary of the incident, 1975, produced a spate of articles. William R. Halliday, in his 1974 book, *American Caves and Caving*, focuses on aspects of Floyd Collins' caving methods that represent unsafe techniques. Halliday views the accident as a prime example of how a cave rescue should *not* be conducted. He states that with today's technology, getting Collins out safely would be a matter of a few hours, or perhaps a day. Probably teams of cavers would find a passage from Mammoth Cave to Sand Cave, and, since they approached Collins from the other side, they could simply lift the rock off his leg. Recent explorations by the Cave Research Foundation from Mammoth Cave reached within a few hundred feet of Sand Cave.

It was inevitable that Hollywood would discover Floyd Collins. In 1951 Paramount Pictures with director Billy Wilder created *Ace in the Hole*, a film about a man trapped in a shallow southwestern cave. An unscrupulous reporter artificially prolongs the rescue to milk the story for all he can get—and to carry on an affair with the trapped man's wife. The film, most successful in depicting the monstrous vulgarity of mob behavior at the scene of a bizarre catastrophe, shows how infecting and corrupting sensationalized news can be. But the film received poor reviews, landing in the archives after a short run to be retrieved years later for late-show television. In 1956 *Ace in the Hole* was the object of a lawsuit, when playwright Victor Desny sued Wilder and Paramount for allegedly stealing his photoplay idea based directly on the Collins story. After much litigation Desny's claim for damages was sustained by the California Supreme Court.

Around 1955 author Robert Penn Warren appeared in the Cave City area, gathering information for his novel, *The Cave*, based on the Collins story. The plot of this rather overdrawn work focuses on the lives of its five narrators, all of whom exploit the situation for profit and fame. It is the local Baptist preacher's son who is the key manipulator of this prolonged rescue.

While neither the film nor the novel attempted to be faithful in detail to the original story, they do share some characteristics of the Collins episode. In all of them the emphasis is on those involved in the rescue rather than on the trapped man; the personal and commercial exploitation of the incident for fame and profit; the morbid nature of the public's curiosity; and the human capacity to use a tragic event for vulgar self-indulgence.

In 1961 a portrayal of the Collins tragedy based on William B. Miller's reporting and supervised by him was presented on the television series "The Big Story," dedicated to dramatizing prize-winning news stories.

But Why Floyd Collins?

Fifty years later the question still remains: Why was Floyd Collins singled out as a national hero for dying imprisoned underground? The two most crucial factors probably were the time span and Collins' condition. The Sand Cave rescue lasted longer than any other comparable effort, allowing the press time to fully exploit the incident to make a lasting impression on the public. The American tradition of "rugged individualism" undoubtedly enhanced Collins' appeal as a hero. Floyd Collins fearlessly caved alone, was trapped alone, and died alone. Miller's extraordinary reporting permitted the public to share intimately in the man's personal ordeal of life and death, and to identify with him. In addition, the nineteen-twenties were times of transition from rural to urban living for many Americans. The Collins tragedy involved simple, fundamentalist country folk with whom many Americans, only one generation off the farm, could identify.

The deep-seated human interest in suffering, dying, and death has always drawn people to the story. There is an undeniable element of horror in Collins' ambiguous circumstance. Miller's descriptions of Sand Cave contain the elements that make up popular "cave-fear"—the dark, the cold, the wet, the silence, the mysterious, and, ultimately, the dangerous. It is easy for one to imagine the hideously agonizing tortures this fellow human being endured before death ended his suffering. In spite of widespread fear of death, there also exists an intense attraction to the phenomenon. Collins in his cave afforded people an opportunity to observe—at a safe distance—a man slowly dying. They could indulge in their private, perhaps unconscious, death wishes in a socially approved manner, and yet hope for his rescue. Vicarious involvement with Collins' confined and deteriorating situation possibly provided many people with a feeling of personal freedom, security, and even well being by comparison. The ultimate appeal of Collins' fate may be the morbid human fascination with the thought of being buried alive. Like a Gothic tale by Edgar Allan Poe, Floyd Collins' story allows people—comfortable in the sunny surface world—a restless squirm of delicious horror.

As the years pass who Floyd Collins really was decreases in significance as he becomes more of an American folklore figure. As with other folk heroes, his stature and legend continue to grow. Recently a reporter passing through Cave City searching for information on Floyd Collins was told by an old-timer enjoying a bottle of soda pop at the gas station: "Don't you pay no attention to any of 'em, Floyd Collins ain't dead at all. They done it to get these here caves in the newspapers. It was just a dummy they took out that sinkhole. Floyd's living on a ranch out in Arizona."

Sloth Droppings

The lack of weathering, the even climate, and freedom from disturbance in caves often preserve animal or human remains that would soon be destroyed on the surface. About 11,000 years ago the Shasta ground sloth suddenly disappeared. Most of what is known about this unusual animal comes from the dung it left behind, preserved in a few remote caves. Its extinction seems to have occurred just as human hunters appeared in large numbers in the western United States. Did man kill off the giant ground sloths?

Paleontologist Paul S. Martin has been investigating the causes of the rapid disappearance of many species of animals during Pleistocene times, and has written widely on the subject. His work has taken him over much of the globe, including all nine caves known to contain sloth dung. Dr. Martin is professor of geoscience at the University of Arizona.

Sloth Droppings

PAUL S. MARTIN

The climb up the steep talus slopes was exhausting. I catch my breath at the gate blocking the entrance to Rampart Cave in northern Arizona. Five hundred feet below, the Colorado River flows through a gap in Grand Wash Cliffs, leaving Grand Canyon and entering Lake Mead.

The National Park Service guides unlock the gate for me and, after briefly inspecting the cave's contents, they return by boat down the lake to the Temple Bar Ranger Station. I watch the winter sun drop behind sheer walls. To the northwest, a growing cloud bank portends a rapid change in weather. Dark shapes emerge from the tamarisk thickets along the Colorado River. Austin Long and six of our students from the University of Arizona are backpacking in from the Pierce ferry landing to join me at the cave. I watch as they thread their way upslope between huge blocks of detached Muav limestone left behind by an ancient mudflow.

Inside the gate is one of the most remarkable fossil deposits in the world. Its value can only be compared with such other

---

Grand Canyon and the Colorado River are outlined by the entrance of Rampart Cave. Ground sloths first entered the cave more than 40,000 years ago, disappearing completely about 11,000 years ago. *National Park Service, courtesy Paul S. Martin*.

paleontological treasures as fossil insects preserved in amber or the frozen mammoths of Siberia. The cave mouth leads into a forty-by-fifty-foot chamber, the floor of which is covered with something even more remarkable than fossil bones. Rampart Cave, one of only nine such cave sites known, holds the biggest and best-preserved deposit of ground sloth dung (and I've seen them all) in the world.

Trenches dug in the deposit by earlier paleontologists exceed four feet in depth. Twigs mixed with the fecal pellets of pack rats are sandwiched between layers of sloth dung. The twigs and other plant material were brought into the cave by pack rats over a period of thousands of years. Beneath vaults in the ceiling lie thin layers of free-tailed bat guano. Scattered bones—of a ground sloth, an extinct yellow-bellied marmot, a Harrington's mountain goat, or a pack rat—and even sloth hair can be found in the layers exposed by previous research excavations. But most of the deposit, nearly 200 cubic yards, is composed of dung of the extinct Shasta ground sloth (*Nothrotheriops shastense*). Untrampled lumps in corners are the size of softballs. Identification is all but certain. No other animal in Arizona, extinct or living, is known with dung of this size and texture.

The dung looks and smells fresh, as though recently dropped. The carbon–nitrogen ratio on the surface of the dung is similar to that of cow manure, further evidence of the quality of preservation. The ancient fecal remains still contain much water-soluble material reminiscent—in color and odor—of a barnyard drain.

The first cave found to contain ground sloth dung—along with hair and even a large piece of hide—was discovered, not in North America, but in the southern tip of South America. At the turn of the century, travelers and scientists converged on a large cave near Ultima Esperanza Sound in southern Chile. The perishable remains in the cave led the famous Argentine paleontologist Carlos Ameghino to describe what he believed was a new species of ground sloth and to predict that living ground sloths would be found in the region. A London newspaper financed an expedition to search for living ground sloths. The scientific world was ablaze with curiosity. The dung balls of the

South American sloths were larger than those of circus elephants!

But no living ground sloths were reported. Another thirty years passed before caves containing sloth dung were discovered in North America. And no ground sloth remains have emerged from geologic deposits of the last 10,000 years. For these reasons, paleontologists have discounted any romantic notions of a late survival, such as the possible persistence of living ground sloths in some remote corner of South America or at the bottom of the Grand Canyon.

Four genera of ground sloths lived in North America during the late Pleistocene, and twice as many genera inhabited South America. Their living relatives, the tree sloths, are diminutive and more highly specialized. Ground sloths may not have been quite as languid and slow moving as their contemporary arboreal relatives, but their stubby, arthritic-looking bones and long, unretractable claws must have made it impossible for them to move quickly. They would have had to endure the attacks of predators while standing their ground. Their anatomy suggests that while some species were browsers, no ground sloth climbed trees.

The largest of the extinct genera, *Megatherium*, apparently was larger than an elephant. The shearing molar cusps of these monstrous terrestrial animals suggest that they were superbly adapted browsers. In the rich tar bed fauna of Talara, Peru, for example, Rufus Churcher of the Royal Ontario Museum discovered sheared twigs that fit the occluding cusps of *Megatherium* teeth from that deposit. Radiocarbon dating of the twigs revealed that they were 14,500 years old.

Another peculiar group of ground sloths, the mylodonts, were almost rhinoceros-sized and possessed a dermal armor formed of closely spaced, peanut-sized bones embedded in their thick hides. Even if other bones are lacking, the presence of the distinctive dermal bones is diagnostic of the presence of mylodont sloths in a fossil fauna.

The largest surviving piece of ground sloth hide, found in the Ultima Esperanza cave, is on display in the Museo de la Plata, Argentina. It is almost three feet in diameter, partly folded, with

Paleontologist Paul S. Martin, respirator securely in place, carefully examines undisturbed ground sloth dung balls. Markings help to identify samples. *Gene Griffen, courtesy Paul S. Martin*.

patches of hair remaining. The embedded dermal bones establish its identity as a mylodont. Radiocarbon dates recently obtained from this prize specimen show that, despite its appearance, it is quite old. The radiocarbon dating method indicates that those in search of living ground sloths in southern Chile came 13,000 years too late.

The Shasta ground sloth belonged to a third group, the megalonychids. This species was the smallest of the North American genera if we exclude a few relatives that found their way into the Greater Antilles. The pony-sized Shasta ground sloth had short legs, a bulky paunch, and may have weighed 300 to 400 pounds. In its Pleistocene heyday—the time of the mammoths and native North American camels and horses—it ranged from northern California and the Texas Panhandle, south into northern Mexico. Except for the famous La Brea tar pits, in what is now downtown Los Angeles, the Shasta ground sloths' fossilized bones are not very common, suggesting that they weren't either.

Almost fifty years ago an articulated Shasta ground sloth skeleton, virtually complete with even the small foot bones held together by dried ligaments and tendons, was found in the bottom of a fumarole on the side of Aden Crater, New Mexico. Speculating on its age, Professor Lull of Yale University found that "its amazing condition of preservation gives one the feeling that it cannot be more than a few hundred years old." History was repeating itself. As in the case of the mylodont hide from Chile, it was hard to dissociate the idea of elegant preservation from the thought of recent age. Nevertheless, Lull assigned the mummified carcass to a much earlier time, an interpretation subsequently supported by radiocarbon dates, which showed that the animal died at least 10,000 years ago.

One aspect of Shasta ground sloth ecology is well known. That is the matter of their diet.

Paleobotanists have long appreciated sloth dung for its value in determining the ancient plant communities of arid regions. Richard M. Hansen of Colorado State University recently examined plant fragments from various levels in the sloth dung deposit of Rampart Cave. He found that the sloths browsed on cat's claw (*Acacia*), Mormon tea (*Ephedra*), globemallow (*Sphaeralcea*), saltbush (*Atriplex*), mesquite (*Prosopis*), and succulents (*Opuntia, Yucca, Agave*). Pollen of Mormon tea and globemallow is much more evident in the dung than in soil samples taken from outside the cave. Near Rampart Cave these plants flower only in late winter or in early spring. Late winter is also the best time for finding green forage at the low elevations in this arid area, and I believe that was the time of year the sloths used Rampart Cave.

Also, from the occasional presence of embryonic sloth bones, I suspect that females used the caves to give birth.

Acting as a coroner's jury, Long and I hoped to determine the cause of the Shasta ground sloth's extinction. Our first need was to determine when the animals of Rampart Cave died. By radiocarbon dating of their dung, we could at least discover when they lived, and we hoped that by judicious sampling from the top of the deposit, we would discover the time in the prehistory of the Grand Canyon when sloths last occupied both Rampart and Muav, an adjacent ground-sloth cave.

From fifteen radiocarbon-dated samples collected at various depths, we determined that the sloths first entered Rampart Cave more than 40,000 years ago. For some reason, they left about 32,000 years ago, abandoning the cave to pack rats.

Twenty thousand years later, the sloths returned to their favored shelter. Between 13,000 and 11,000 years ago, they flourished, judging by the quantity of dung deposited during that interval. The youngest of our dung samples was 10,780 years old. About that time, the Shasta ground sloth seems to have died out suddenly, not only at Rampart Cave but at similar caves elsewhere.

I doubt that the Shasta ground sloths spent much time in the cave. Had they done so, they would have soon filled it with dung. We estimated that the average annual rate of deposition was only slightly more than one cubic foot a year, an amount that probably represented less than a week's elimination from one healthy adult ground sloth.

The summer season is oven hot at the bottom of the Grand Canyon. July and August rains, which renew plant growth in other parts of Arizona, rarely fall along the Grand Wash Cliffs. In the absence of tender new browse, I believe the sloths left the canyon in spring, to summer at higher and cooler elevations. At 5,000 feet, site of the present Hualapai Indian Reservation, they would have found cooler temperatures and Joshua tree–juniper–sagebrush woodland. (From dung studies in Texas and Nevada caves, we know that the sloths browsed on woodland trees and shrubs.) The vertical migration from canyon bottom to the woodland area would have taken no more than a few weeks at a possible sloth travel rate of a few miles per day.

Various dietary studies that have been conducted on the fossil dung of the Shasta ground sloth show that its favorite food plants—in the Rampart and Muav cave areas, Gypsum Cave, Nevada, the sloth caves of the Guadalupe Mountains of west Texas, and Aden Crater, New Mexico—are still important components in the vegetation of arid regions in North America.

Thanks especially to the remarkably stratified record of Shasta ground sloth diet available from the dung deposits at Rampart Cave, we know that for thousands of years they browsed on a variety of desert and woodland shrubs, including species presently favored by wild desert bighorn sheep and feral burros. Over thirty genera of plants identified in sloth dung remain important in the natural vegetation at or near the Grand Canyon sloth caves. It is not easy to account for the Shasta ground sloth's extinction by loss of food supply.

Why, then, did the sloths vanish? That is the question Long and I were pondering on our winter's trip to Rampart Cave. Although I discount the idea, one cannot exclude some drastic and sudden climatic catastrophe, perhaps one that briefly blighted the land, leaving no trace of its occurrence. I have more faith in another possibility, no less catastrophic in terms of the sloth's ability to survive.

Archaeologists have determined that 11,200 years ago big-game hunters pursued mammoths along the San Pedro Valley of southern Arizona. These hunters seem to have suddenly appeared in the western United States at this time. Admittedly, Paleo-Indian artifacts, including the Clovis spear points that are occasionally associated with mammoth bones, have not been found in association with the ground sloths. In fact, no archaeological material has ever been found in convincing association with sloth bones or dung. But lack of kill sites is not necessarily proof against the overkill theory, which holds that the massive extinction of the New World Pleistocene megafauna was caused by human hunters in a relatively short period of time. Abundant kill sites may never be found, and the true nature of the human impact may not be recorded in our fossil record—which is all too incomplete as paleontologists commonly complain.

But the fact remains that the sloths died out just at the time when, according to archaeological evidence, the first big-game hunters arrived in North America. Slow, lumbering, and leaving large and distinctive droppings, the ground sloths would have been easy to track and kill. There is no reason to believe that the first hunters could perceive, much less control, the impact of

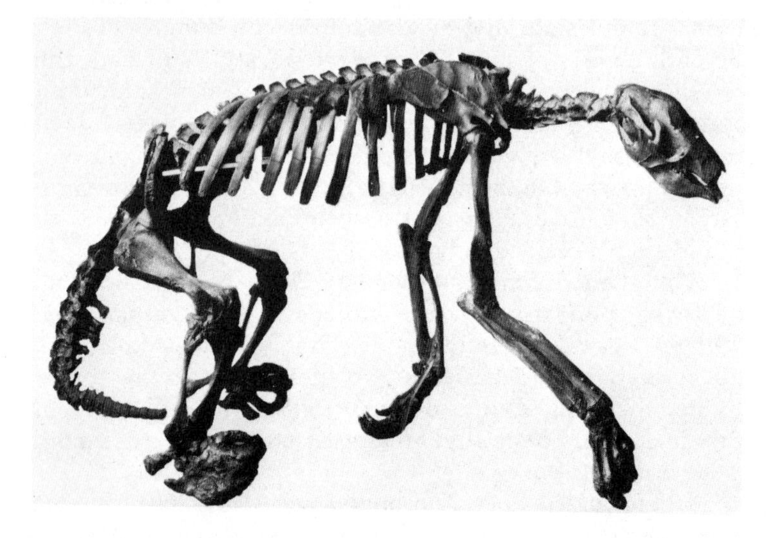

This restored skeleton of a Shasta ground sloth is now at Peabody Museum, Yale University. This species, among the smallest of the ground sloths, was pony-sized and weighed 300 to 400 pounds; others were as large as elephants. The Shasta ground sloth—extinct for some 10,000 years—was a Pleistocene contemporary of mammoths and native North American camels and horses. *Paul S. Martin*.

their arrival on the more vulnerable and easily destroyed native large mammals of North America. Enjoying an abundant food supply and not threatened by any serious enemies themselves, the hunters could have increased very rapidly.

I do not exclude the possibility of a rate of increase as fast as a doubling in numbers per generation. Within a few hundred years the human predators could have swept over most of North America, obliterating much of the big game.

A front of highly skilled hunters could have swept through any one region in a few years, leaving little evidence for archaeologists to fish out of the fragmentary fossil record.

If I am right, and I admit that other possible explanations for sloth extinction exist, the best evidence for man's presence in any region may, paradoxically, be the absence of ground sloths and other large, vulnerable prey. The youngest dates on ground-sloth middens may say as much about human culture as the oldest dates on human artifacts.

For this reason, I hope anthropologists will join paleontologists in treasuring the scientific significance of ground-sloth dung. From the dung we have an opportunity to study both the ecology and the natural history of a remarkable group of animals.

Because the caves containing sloth dung are not numerous, and the contents perishable, their value is accordingly great. The sloth caves deserve the same protection as that afforded ancient ruins of prehistoric peoples. We have much more to learn about a mysterious moment, about 11,000 years ago when, for the last time, the dung came down.

STEPHEN BISHOP: The Man and the Legend

Visitors to Mammoth Cave in Kentucky increased greatly following the War of 1812. The cave was soon world famous, and attracted scientists, journalists, and people of renown who expected to be conducted through the cave by knowledgeable competent guides. The early guides to Mammoth Cave were slaves, and Stephen Bishop was one of them. His unusual talents made him so widely known that for a while he was almost as famous as the cave itself. The stories about him and his feats survive. With their telling and retelling, the legend of Stephen Bishop was born—and still survives to this day.

THE RIVER STYX

STEPHEN BISHOP: The Man and the Legend

HAROLD MELOY

Mammoth Cave has attracted millions of visitors during its more than 175 years of recorded history. Generations of guides have conducted cave tourists along its miles of underground avenues. Some of the guides became popular with travelers and writers, but the most celebrated of all was Stephen L. Bishop (1821-1857), who became almost as well known as the cave itself during the mid-nineteenth century.

Stephen was a slave; yet he became a legend during his own lifetime. Sought after by authors, scientists, and journalists to guide them through Mammoth Cave, he conducted them through the cave corridors with dignity and confidence.

The cave had become famous before Stephen was born. White men came to Mammoth Cave during the 1790s to dig its saltpeter, which they used to make gunpowder. Pioneers and adventurers visited the cave during the first decade of the 1800s. A published description by one of them appeared in 1810. Cave maps were made or copied from older maps in 1811, 1813, and

---

River Styx. Illustration from *Pictorial Guide to the Mammoth Cave* by the Reverend Horace Martin, Stringer and Townsend, New York, 1851.

1815, some of the maps purporting to show twenty miles of passages. Saltpeter mining and processing continued until early 1815.

An Indian mummy discovered in nearby Short Cave was brought to Mammoth Cave for exhibition in 1813. Soon the number of newspaper and magazine articles publicizing the cave increased—and so did the number of visitors. The mummy was taken away in 1815, but after Stephen became a guide in 1838, he retold its story so vividly that it became one of the traditions of Mammoth Cave.

Prior to 1815 the mining superintendents—Archibald Miller, Fleming Gatewood, and John Holton—doubled as guides for occasional visitors. Miller and Holton continued until they were replaced by a second generation of guides. George S. Gatewood, son of Fleming, began guiding in 1827. Other guides during the next ten years included his brother E. B. Gatewood, Robert S. Bell, Robinson Shackelford, his son Joseph C. Shackelford, and Archibald Miller, Jr., the son of the former mining superintendent. George Gatewood made numerous explorations in the cave during the early 1830s. Stephen later was to find Gatewood's name in many of the most remote parts of the cave, parts never shown to tourists.

During the winter months of 1834-1835 Edmund F. Lee, a civil engineer from Cincinnati, made a complete instrument survey of the cave. George Gatewood was his guide and almost constant companion, telling him the names of the passages and principal places of interest. According to Lee's measurements, the length of all of the cave avenues totaled about eight miles. After his survey was completed, Lee prepared a beautiful scale map of the cave and published a thirty-page guidebook, *Notes on the Mammoth Cave*. This was the standard reference and guidebook used when Stephen came to the cave three years later.

Franklin Gorin, an attorney in Glasgow, Kentucky, had known about the cave all his life. Lee's guidebook heightened his interest. On April 17, 1838, Gorin and A. A. Harvey signed promissory notes to purchase Mammoth Cave for the sum of $5,000. The purchase price was to be paid in annual installments during the next five years. Gorin enlarged and renovated the visitors' accommodations. Although it didn't appear so at the time, Gorin's most significant contribution was bringing his young slave, seventeen-year-old Stephen, to be a new guide at Mammoth Cave.

Few pictures of Stephen Bishop are available. This one was engraved by N. Dmitrieff, and first published in *Scribner's Monthly Magazine*, October, 1880, with an article by Horace C. Hovey.

The guides Archibald Miller, Jr., and Joseph C. Shackelford taught young Stephen the routes and passages, as they had been taught by their fathers. They told him the names of the avenues and the points of interest as set out in Lee's guidebook, and they related the stories that visitors enjoyed hearing. They showed Stephen the saltpeter hoppers used during the War of 1812, and the road through the main cave along which oxen had pulled carts hauling the ore to the hoppers. Tracks of the cartwheels remained in the road, as did the corncobs at the place where the oxen had been fed. Many of the log water pipes were still in place. Others had been moved and arranged as seats in a spacious cave room used for church services during the previous five years.

Stephen climbed the ladder to the Haunted Chambers. The guides pointed to a ledge where they told him the Indian mummy had been found. The mummy was no longer there. According to them it had been taken by a Mr. Ward to a museum in Massachusetts. Legend had already proclaimed her an ancient queen, and legend had provided a baby mummy found at her side, each sitting in adjoining niches at "the mummy seat."

Remnants of cane torches and discarded Indian moccasins were found in abundance along the passages as far as two miles from the entrance. The guides told him that the early miners had found many skeletons, one of gigantic size, while digging for saltpeter. Stephen was thrilled where others shuddered at the deep pits within the cave. Crevice Pit was reported to be 280 feet deep. They did not know the depth of Bottomless Pit (a name taken from the Bible). A whole new world opened to the new guide, and he was to make it his own.

Before long Stephen was guiding visitors along the tourist trails through the cave. One of them was a young man known to Stephen as Wandering Willie, who had walked from his home in Cincinnati to Mammoth Cave, carrying only his violin. He played it in the cave. Stephen thought the music was beautiful. That evening Willie asked permission to pass the night in the cave. He promised to remain at the same place until the guide called for him the following morning. Willie selected for his underground bedroom a place at a spring in the main corridor of the cave.

The following morning when Stephen awakened him, Willie told his new friends that he had spent a glorious night in the cave alone with his violin. Such is the stuff from which legends are made. From that time on, the spring where he slept was known as "Wandering Willie's Spring." Stephen was fond of telling the story to the cave visitors; some of the visitors were authors, and they retold it in their books.

One day that summer, some of Gorin's friends and relatives visited his cave. His nephew Charles Harvey became separated from the others, took the wrong turn in one of the passages, and was soon lost. When he was missed an immediate search was begun. All available men in the area were summoned to help find the missing man.

Young Stephen joined the others and without hesitation entered lesser known passages off the tourist trails where he had never been before. After thirty-nine hours Harvey was found in a side passage, since known as "Harvey's Lost Way." Stephen had seen many parts of the cave that were new to him. He was enchanted.

He familiarized himself with every passage known by two generations of previous guides. More than that, he began to blaze new paths in the darkness where others had never gone before. Some were through confusing mazes, vertical as well as horizontal. One of these was a narrow, twisting, canyon-like passage to the side of and below the path to Bottomless Pit. He scaled one wall, descended another, and continued through the labyrinth until a window opened in the left wall. Beyond the window was a tremendous vertical void, the largest domepit yet found. Gorin gave it his own name. Ladders were placed in the "labyrinth" leading to Gorin's Dome so that visitors could more easily reach this new-found wonder.

Newspaper descriptions of the new discovery brought adventurers eager to join in further explorations. What lay beyond Bottomless Pit? The guides could see that the passage continued beyond, but the pit was so wide and deep that explorers during the previous forty years had been unable to cross it. With their feeble lights they could not even see the bottom. On October 20, 1838, a Mr. H. C. Stevenson of Georgetown, Kentucky, and Stephen stood on the brink of Bottomless Pit. They could see the passage continue beyond, and they planned to reach it. Would the ladders in the labyrinth reach across the pit? Yes, they would. Carefully they placed a ladder across the pit. According to an 1842 French publication, Stephen then placed a second and longer ladder beside the first. Carefully they crossed over to the other side. The cave went on and on; and so did they, as far as the oil for their lamps permitted. Then they retraced their steps to the entrance and up the hill to the inn.

Gorin was elated with the new discovery. No time was lost in spanning the pit with a wooden bridge. Guides and visitors alike used the bridge for further explorations. Within days Stephen had explored miles of new passages beyond the pit. Descending to lower levels, he found an underground river, then another, and another.

Lee's map and guidebook, only three years old, were now woefully out of date, for Stephen and his companions had found as much cave beyond the pit as was known before. Reports of these discoveries brought an increased number of visitors, and more guides were needed to accommodate them. In 1839 Gorin hired two slaves of Thomas Bransford of Nashville, Tennessee, for the annual sum of $100 each. The new guides were the brothers Mat and Nick. As Stephen taught them the cave, they caught his contagious enthusiasm for further exploration.

In October 1839 Stephen met his new master. Gorin had sold Mammoth Cave to Dr. John Croghan of Louisville for the sum of $10,000. Stephen was included in the deal. Mat and Nick continued as guides under the same arrangements with Bransford; Alfred, another slave, was added to the guide staff. Explorations continued and new discoveries followed, one after another. In November 1840 Stephen and a visitor with a zest for strenuous caving climbed a passage wall and entered a hole near the ceiling. They followed a long muddy stoopway until it ended at another domepit. This was even larger than Gorin's Dome. Another major discovery had been made. The next month, a larger party returned with ropes. Led by Stephen and Mat, they descended the pit and found a new cave chamber which surpassed all their expectations. None of them had ever seen anything like it before. It was mammoth, and Mammoth Dome was added to the itinerary to awe the cavers who followed.

The rivers were a challenge. The largest had been called the Jordan River, but after Stephen heard echoes along its course, he renamed it the Echo River. Only a small boat could be carried through tight passages to the river's bank, and then only when the water level was low. But the waters were unpredictable. Sometimes the river would flood above the approaches to the river passages. The first boat was lost by rising waters, another was smashed. They finally crossed Echo River in 1840, and Stephen entered the many branching passages beyond.

Stephen's reputation as the principal cave guide increased with each new discovery. He was fearless and dependable. Visitors who came for exploration insisted that they have him as their guide. In July 1841 John Craig of Philadelphia and Brice Patton of Louisville asked to be taken to the farthest point in the cave. Stephen led them through a corridor far beyond the rivers. Working their way up into a higher cross passage, they discovered another avenue containing many strange and beautiful white gypsum flowers.

In one area the ceiling was covered with white nodules that reminded them of snowballs thrown against it. At other places the gypsum formations were more delicate and fragile, glistening and sparkling in the illumination of their cave lights. This was another major discovery. The new avenue was christened Cleaveland's Cabinet after Professor Parker Cleaveland's Cabinet of Crystals at Bowdoin College, Brunswick, Maine. Stephen delighted in showing these treasures to scientist visitors. He listened intently to their discourses and explanations. He remembered and used the same language with the same Greek and Latin terms for the benefit of later tourists, who were amazed at his knowledge of Greek and Latin and his grasp of the principles of geology.

For some time certain doctors in Louisville had felt that the constant cave temperature and atmosphere would be beneficial for treatment of their consumptive (tubercular) patients. Some of them encouraged Croghan to establish a sanatorium in the cave, and he agreed. In 1841 he erected several frame cottages in Audubon Avenue within half a mile of the cave entrance. A consumptive patient had planned to arrive that year but was too weak to travel to the cave. It was the following year before the first patients arrived for the unique underground medical experiment.

During the winter of 1841-1842, when there was little if any business at the cave, Croghan took Stephen to his home near Louisville. John's brother, Colonel George Croghan, was also there. The colonel used the opportunity to have Stephen prepare a new and up-to-date map of Mammoth Cave. Lee's map was used as a guide for the eight miles of cave passages this side of Bottomless Pit. Stephen penciled in perhaps twice that many miles of passages he had discovered beyond the pit. George inked the penciled lines and added the names of the chambers, rivers, and passages as Stephen named them. By watching George letter the map, Stephen received one of his first lessons in reading and writing. The map was completed by the end of January 1842. The Croghans gave Stephen full credit for preparing the new map. Copies were made to send to relatives and for use at the cave. Three years later the map was published as part of a new book about the cave.

Louisville had other attractions for Stephen. He met and married a pretty mulatto girl, the eighteen-year-old Charlotte, also belonging to Croghan. Charlotte accompanied them back to the cave in the spring of 1842. She became a maid in the Mammoth Cave Hotel which Croghan was again enlarging.

Croghan expected several consumptive patients and he was at the cave to meet them. First to arrive was Dr. William J. Mitchell, himself a medical man, who had diagnosed his own ailment as pulmonary consumption. On May 23, 1842, Dr. Mitchell took up residence in one of the cottages in Audubon Avenue. At the end of five weeks he pronounced himself "very much relieved" and left the cave. This apparent success was welcome news to other doctors and their patients. Later, John Wesley Harper from South Carolina became a patient. He was followed by Oliver H. P. Anderson. Anderson first occupied a cottage near the Star Chamber; later Croghan erected a cottage for him in Pensacola Avenue. Some of the other patients were Benjamin R. Mitchell of Mobile, Alabama; Reverend Charles Marshall from Oswego County, New York; and Oliver P. Blair of Pittsburgh.

Stephen knew them all. He would sometimes walk the cave trails with Anderson; on December 23, 1842, at the far end of Pensacola Avenue, Anderson wrote their names: "O. H. P. Anderson" and "Stephen L. Bishop." Anderson left the cave January 11, 1843. Those with terminal illnesses died in the cave, and the experiment failed.

In the meantime Dr. John Locke, a geologist from Cincinnati, came to study the cave. Stephen showed him around and helped him take barometric measurements of the five different levels of the cave. He learned from Dr. Locke more scientific language and more Greek and Latin words. Each was learning from the other. Locke planned to write a book describing the geology of the cave; Stephen repeated Locke's geology to later visitors.

One of the stone huts erected in 1842 for the use of consumptive patients. *Pete Lindsley*.

Since his earliest trip into the cave, Stephen had seen names written on the walls and ceilings. Some of the names were scratched or carved in the stone walls; most of those on the ceilings had been smoked by candles or lamps. Many of the names were accompanied by dates; the earliest date was 1801. Cave visitors guided by Stephen often read the old names to him; and if he asked them, they would spell them, naming each letter. They added their own names, as Stephen watched, and sometimes added his name too. Stephen was an eager student. After a few months of such instruction, he had learned to read and write. Now he could place his own name at the most remote parts of the cave. Sometimes he signed it "Stephen L. Bishop." No one to this day knows his middle name.

Near the far end of beautiful Cleaveland Avenue was an even more beautiful branching passage, profusely decorated with delicate white gypsum flowers. He named it Charlotte's Grotto for his wife. Sometimes at night, when their day's work was completed, he would take her into *his* cave. One night at the far end of the Black Chambers he wrote on the wall "Charlotte Bishop, Louisville, 1843." On the wall of another avenue he

The name "Stephen L. Bishop" on the wall in Mammoth Cave. *Pete Lindsley*.

wrote "Stephen L. Bishop and Mrs. Charlotte Bishop." Their son Thomas was born that year.

More and more people came to see the cave. A number of them were journalists who wrote newspaper and magazine stories about it—and about Stephen. A new sixty-eight-page Mammoth Cave book by William Basil Jones was published in 1844. Although it included a number of the discoveries made by Stephen and described all of the cave then being shown, its language was too grandiose, flowery, and exaggerated to meet with Croghan's fastidious taste; he did not promote the sale of the book.

Besides, Croghan had his own ideas for a new cave book. His old friend the journalist Alexander Clark Bullitt visited the cave that year. Stephen took Bullitt on all the tourist routes. He told him the legends that had been handed down from previous guides and the more recent events in which Stephen had taken a role. These would have been sufficient material for a book, but Bullitt went further. From the wealth of material that had already been written, he selected excerpts from six prior publications to weave into his narrative. Croghan approved the completed manuscript and published it in 1845 under the title *Rambles in the Mammoth Cave During the Year 1844 by a Visitor.* The completed book included a fourteen-by-twenty-inch lithograph of Stephen's 1842 cave map. Bold letters on the map proclaimed that it had been prepared "by Stephen Bishop, one of the guides." His reputation soared.

The great Norwegian violinist Ole Bull visited the cave in 1845. Stephen conducted him, with Louisville publisher George D. Prentice and Mrs. Prentice, beyond the rivers to the end of Silliman's Avenue. There, in a spacious hall, the renowned artist played the violin. "The reverberations were fine ... the music had a weird, unearthly sound, as it was echoed through those eerie, uncanny chambers." Stephen later related that the great artist "seemed very much excited and gave him new ideas of the agility of music." After the foursome left the cave, Ole Bull was still excited. He and Stephen retraced their steps back into the cave, and there he played again. Stephen called the room "Ole Bull's Concert Hall." Other visitors agreed that the wild and rugged grandeur of the chamber matched the tempestuous style of the celebrated concert violinist. Ole Bull went on to other triumphs, and in his underground concert hall Stephen told and retold the story to visitors.

While the cave and Stephen were growing in reputation, the Mammoth Cave Hotel was growing in size and hospitality. To the inn enlarged by Gorin, Croghan added two log buildings, each with four rooms, a twelve-foot hall, and a second story. Later, a larger two-story building thirty by ninety feet was added. The first floor of this building was the dining room, the second floor the ballroom. During the summer months Croghan brought musicians from the Louisville theaters to the cave. The hotel orchestra played during dinner in the dining room, and in the evenings for dancing in the ballroom. During the day, the musicians often earned extra money by accompanying the more affluent guests into the cave. They played one type of music for the majestic underground avenues, another for the magical Star Chamber, and still another for the boat rides on Echo River.

More and more wealthy southern families were coming to Mammoth Cave during the hot summer months. Croghan built sixteen cottages as an extension of his hotel for their use. During the hot days and nights in the summer of 1846, the young people proposed that a dance floor be built within the cave. Local carpenters constructed the floor, the musicians played, and the guests danced gaily in the underground ballroom. Stephen added the story to his spiel for later cave tourists, and mapmakers added it to their maps.

Editor Thomas Kite and his party arrived at the Mammoth Cave Hotel on May 30, 1847. He requested Stephen for their guide. The first day they went to Gothic Avenue. All morning Stephen had been suffering with a severe toothache; he apologized for his condition and said that he "could not make the cave interesting for them." He asked them to wait at Lovers' Leap until he could send another guide. Stephen sent Alfred, who guided them through the rest of that trip through the Star Chamber and on to Chief City. From Alfred they learned more about Stephen and his remarkable affinity for the cave.

The following morning Kite inquired of Stephen's condition; it was improved. During the following days Stephen conducted them through the rest of the cave. Between the rivers, where the water was too shallow for boats, Stephen picked the young ladies up in his arms and carried them to dry land. He ferried the three men in a similar fashion. Arriving at Echo River they embarked in a "long narrow boat." Stephen gave them his demonstration of the echoes along the river. Describing their last trip across the river, Kite wrote in his personal journal: "Gently gliding along . . . the deep quietness of the place broken but by our voices, combined with a feeling that we were indeed saying farewell to it, had rather a stilling effect, and we knew not how long the silence that ensued would have continued, had not Stephen commenced singing 'Home Sweet Home.'"

Landscape painters often came to Mammoth Cave to sketch the underground scenery. In July 1842 "Mr. Allen and Mr. Campbell, artists of great merit, had been for some time engaged in taking views of the cave." Six of these were published in the *Rambles* book. During May 1848, George St. P. Brewer and John Brewer made sketches in the cave for additions to their Moving Panorama in Louisville. Stephen illuminated the scenes with every lighting device known at the time. Over forty different views of Mammoth Cave were included on the oil canvases in Louisville. They were added to views of Niagara Falls and other natural wonders on the 25,000-foot canvas. It is reported that 10,000 feet of the Panorama were devoted to views of Mammoth Cave. The finished Panorama was transported and exhibited throughout the United States, and also toured England for nine years. The paintings were eventually lost.

Different editions of the souvenir book to accompany Brewer's Panorama were printed in Boston, Philadelphia, and London. The text describing Mammoth Cave was taken largely from *Rambles*. An additional paragraph gave the Brewers' feelings about their guide:

> And now let us say a word about the handsome, good-humored, intelligent Stephen, the most complete of guides, the presiding genius of this territory. He is a

> middling-sized mulatto, "owned," as they say here, of handsome, bright features. He has occupied himself so frequently, especially in the fall of the year, in exploring the various passages of the cavern, that there is now no living being who knows it so well. The discoveries made have been the result of his courage, intelligence, and untiring zeal. He is extremely attentive and polite, particularly so to the ladies, and he runs over what he has to say with such ease and readiness, and mingles his statement of facts with such lofty language, that all classes, male and female, listen with respect, and involuntarily smile at his remark. His business as guide has brought him so often in contact with the intellectual and scientific, that he has become acquainted with every geological specimen in the cave, and, having a prodigious memory, has at his tongue's end every incident of interest that has transpired during his administration. Long life to Stephen!

Although he did not mention it in public, Stephen probably knew the identity of his white father. Mulatto slaves were commonplace, but Stephen was certainly not common. In the cave he was a genius and a real showman. When asked by visitors about his parentage, he told them that his father was an Indian. All cave visitors remembered for years afterward not only the skill with which he showed them the cave but also the unusual blood lines of their unique guide.

One jovial party of three that Stephen led through the cave consisted of Robert B. Roosevelt, writer for *The Knickerbocker*, a New York magazine, his friend Tom Wilson, and an Irishman named McCarlin, who had asked to join them. Stephen not only carried extra oil and lights, but "was heavily, although not unwillingly, burthened with comestibles and portables innumerable." Judging from Roosevelt's article in *The Knickerbocker*, they had a rollicking good time. On Echo River Stephen "poured forth, in his deep rich voice, one of the wild songs of his Indian fathers. The tones rang clear and strong, and were echoed and re-echoed back. High would the notes swell, and ring off into the hidden caverns, and then sink so low as to be scarce heard, while the rushing echo of the first would come rolling back—an answer from another and unseen world."

Roosevelt explained to his readers that Stephen "was a slave, his mother of the African species, and his father an Indian, who had learned to read and write by seeing gentlemen paint their names with smoke on the cave walls, and then asking how they spelled them."

Dr. Croghan died on January 11, 1849. By his will he devised Mammoth Cave in trust for the benefit of his nephews and nieces. The will also provided that his slaves be hired out for a period of seven years and then emancipated. The Inventory and Appraisement of his estate, under date of August 13, 1849, valued Stephen, age twenty-eight, at $600, Charlotte, age twenty-six, at $450, and Th. Bishop, age six, at $100. Stephen now belonged to the Trust, as well as to the cave.

In June of 1850 Stephen watched as a special party of seventeen arrived at the hotel. It did not take him long to learn that their leader was Ralph Waldo Emerson, one of the most conspicuous and popular lecturers on the American scene. They came equipped with Roman candles to better light the cave. Stephen took along an ample supply of Bengal lights in addition to the hard-oil lamps for each person. They stayed fourteen hours in the cave the first day. On the second day Stephen led them past Giant's Coffin and on to the Star Chamber where "the guide took their lights and hid them."

Emerson wrote his wife that then "you find yourself at once under a starry sky, with a comet, too, easily distinguishable. The illusion is perfect. I lay there on my back on the ground for a quarter of an hour or more whilst our choir sung 'The Stars are in the Quiet Sky,' and considered that this was the best thing in the cave, and that this was an illusion!" Emerson echoed his feelings in his essay "Illusions," published ten years later.

More to Stephen's liking was Dr. Benjamin Silliman, Jr., a Yale University scientist, who visited the cave the following October. Stephen guided him beyond the rivers and through the long avenue named by Croghan for Silliman's father. Stephen listened with rapt attention as Dr. Silliman explained the

THE STAR CHAMBER

Star Chamber. Illustration from *Pictorial Guide to the Mammoth Cave* by the Reverend Horace Martin, Stringer and Townsend, New York, 1851.

mineralogical aspects of the cave. More geology followed and Stephen absorbed it, word for word. They descended to the bottom of Gorin's Dome. There Silliman erected a Drummond light "and by its aid obtained the first view of its lofty ceiling." The Drummond light, also called the limelight, was the most brilliant man-made light of the time. Often used in theaters, it provided an even more dramatic illumination in the cave—and for a highly appreciative audience.

There was great excitement at the hotel one April morning in 1851. Jenny Lind, the "Swedish Nightingale," and her party were expected. Stephen heard that she had sung over sixty concerts in America. The "Jenny Lind mania" was at its height. In some cities tickets sold for as much as $175, and hundreds failed to find standing room at $3.00 each. Many of her appearances grossed over $5,000 each, a huge sum in those days. Newspapers daily reported her every movement. George Croghan, nephew of Dr. John, was on hand to welcome her to Mammoth Cave.

The celebrated singer and her party arrived at noon. Stephen, of course, was to be their guide. He led them to Gothic Chamber, across Bottomless Pit, through Fat Man's Misery, and on to the rivers. Unfortunately, the rivers were too high to cross. On their return, they visited Gorin's Dome, again passed Giant's Coffin, and on to the Star Chamber. "Here the guide shone in all his glory."

The ceiling of the chamber, far above, appeared to be the canopy of the midnight sky. By the dim light of the cave lamps a multitude of gypsum crystals glittered and twinkled like stars in the firmament. Some were large and shone very brightly, others were small and twinkled softly. Stephen then took all of their lamps, and, without telling them what he was going to do or where he was going, suddenly disappeared. The stars went out one by one until the visitors were left in silent darkness.

Unknown to them Stephen entered a lower parallel passage that rejoined the main cave some distance away. The silence was broken by his voice, far off in the cave, saying, "Let there be light." Each visitor turned in the direction of the sound, and to his surprise saw a few rays of light darting through the thick darkness. The light increased, giving the appearance of early dawn. As the light rose higher and higher, the darkness gave way until the cluster of burning lamps in the hands of Stephen was seen.

Suddenly the notes of a violin were heard breaking on the stillness, and Agathe's Prayer from *Der Freishütz* poured its melody on the chamber. Then the party remembered that their violinist Joseph Burke had brought his instrument with him. It was a rare and delicate moment. Thirteen years earlier Jenny Lind had made her debut in the role of Agathe in Weber's *Der Freishütz*. Jenny Lind did not sing in the cave. The newspapers reported that she was asked to do so, but declined. No matter. The mythmakers wove legends of her singing on Echo River, in Cleaveland Avenue beyond the rivers, at the Arm Chair which since bears her name in Gothic Avenue, and at the Rotunda. Later tourists liked to hear such stories, and other guides were glad to oblige them.

For six years the *Rambles* book containing Stephen's map had been the standard reference book about the cave. It was replaced in 1851 by Rev. Horace Martin's *Pictorial Guide to the Mammoth Cave*. This remained the principal guidebook during the remainder of Stephen's life; it was in turn replaced by Charles W. Wright's 1858 guidebook.

Little notice was taken when Nathaniel Parker Willis arrived at the cave hotel in June 1852. Willis was an author and co-editor of the New York *Home Journal*, but he probably saw no reason to mention it. Instead he stayed at the hotel and took a twelve-hour cave tour as an ordinary good-humored traveler—with a twinkle in his eye. Stephen liked him at once, and he liked Stephen. When Willis's description of the cave was published later that year, it surpassed everything ever written about the cave; it still remains one of the three all-time favorites in Mammoth Cave literature. About Stephen he wrote:

> The remarkable Stephen is a slave, part mulatto and part Indian, but with more of the physiognomy of a Spaniard—his masses of black hair curling slightly and gracefully, and his long mustache giving quite a Castillian

> air to his dark skin. He is middle size, but built for an athlete—with broad chest and shoulders, narrow hips, and legs slightly bowed, and he is famous for the dexterity and bodily strength which are very necessary to his vocation.
>
> The cave is a wonder which draws good society, and Stephen shows that he is used to it. His intelligent face is assured and tranquil, and his manners particularly quiet—and he talks to charming ladies with the air of a man who is accustomed to their good will, and attentive listening. The dress of the renowned guide is adapted to dark places and rough work. He wears a chocolate-coloured slouched hat, a green jacket and striped trousers, and evidently takes no thought of his appearance.
>
> He is married. His wife is the pretty mulatto chambermaid of the hotel. He has one boy, takes a newspaper, studies geology, and means to go to Liberia as soon as he can buy his wife, child, and self from his present master. He has tact, talent, and good address.

Willis was told about a bridegroom touring the cave beyond the rivers who fainted from exhaustion. "The famous guide, Stephen, actually brought him back, six miles, in his arms; though, considering the ladders to go up and down, the holes to creep through, the craigs to climb, the rivers and lakes to navigate, the slippery abysses to edge around, and the long passages in which it is impossible to walk upright, it was considered almost a miracle."

Stephen reached the high point of his career by the mid-1850s. Though a slave, he had risen above slavery through his own achievements. Newspapers, magazines, and books gave him national acclaim. Knowledgeable cave visitors sought his services as their guide; and they, as well as those who chanced to be in his tours, praised his manners and decorum. The most reserved and genteel persons trusted their safety to his care. By this time he was receiving wages to prepare him for his legal freedom in 1856, pursuant to the provisions of Croghan's will. Stephen had become a living legend—a legend which was to be altered, changed, and idealized in the years to come.

World traveler Bayard Taylor visited the cave in May 1855. An author of some note and correspondent for the New York *Daily Tribune*, he was a keen observer of the human scene and a factual reporter in polished prose of the places he visited and the people he met. Taylor wrote of the famous guide:

> Stephen, who has had a share in all the principal explorations and discoveries, is almost as widely known as the cave itself. He is a slight, graceful, and very handsome mulatto of about thirty-five years of age, with perfectly regular and clearly chiselled features, a keen, dark eye, and glossy hair and moustache. He is the model of a guide—quick, daring, enthusiastic, persevering, with a lively appreciation of the wonders he shows, and a degree of intelligence unusual in one of his class. He has a smattering of Greek mythology, a good idea of geography, history, and a limited range of literature, and a familiarity with geological technology which astonished me.
>
> He will discourse upon the various formations in the cave as fluently as Professor Silliman himself. His memory is wonderfully retentive, and he never hears a telling expression without treasuring it up for later use. In this way his mind has become the repository of a great variety of opinions and comparisons, which he has sagacity enough to collate and arrange, and he rarely confuses or misplaces his material. I think no one can travel under his guidance without being interested in the man, and associating him in memory with the realm over which he is chief ruler.
>
> Stephen and Alfred belonged to Dr. Croghan, the late owner of the cave, and are to be manumitted in another year, with a number of other slaves. They are now receiving wages, in order to enable them to begin freedom with a little capital, in Liberia, their destined home. Stephen, I hear, has commenced the perusal of Blackstone, with a view to practise law there, but from his questions concerning the geography of the country, I foresee that his tastes will lead him to become one of its explorers.

Stephen became a free man the following year, but he chose

to remain at the cave of which he was so much a part.

Carlton H. Rogers and his party arrived at the cave in May 1856. He would be satisfied with nothing but the best and made arrangements for Stephen to be their guide. In a private publication for his friends, he related their experiences with Stephen on Echo River:

> At the further end of the passage is the famous Echo River, which during high water is merely a continuance of Lethe. The entrance to this river is through an arched gateway of rocks above. We were fortunate in finding sufficient space to admit our boat although for a few moments we were compelled to humble ourselves in a manner neither graceful nor agreeable.
>
> This ordeal passed, we emerged into the open river. While here, our guide fired a pistol, the report of which was deafening. The sound reverberated and echoed from arch to arch, and dome to dome, like continuous thunder. The echo is truly wonderful, and answers fully the descriptions that have been given of it, continuing, I should judge, for some twelve or fifteen seconds. At first, it is remarkably clear and distinct, but changes to a soft and musical cadence as it dies away in the distance. I notified Stephen the night before, when I engaged him to go with us into the cave, that he must be prepared to do justice to his reputation as a singer, as we should expect him to discourse most eloquent music on the occasion.
>
> It is a fact known to all who are familiar with this part of the cave, that sound is very much modified and softened by being produced here; while, at the same time, the volume is greatly increased. The harshest notes become quite mellow, and the most unmelodious voices comparatively sweet. It is not strange then, that Stephen, who has a rich musical voice, and a thorough knowledge of the acoustics in its connexion with this place, should prove an attractive feature in the scene. He sang for us several popular airs appropriate to the occasion—such as "The Canadian Boat Song," "My Old Kentucky Home," and "Oft in the Stilly Night." He would pause long enough at the end of every

> line, for the last word to be echoed back, the effect of which was indescribably fine.
>
> In the song of "Oft in the Stilly Night," the word *night* came back clear and distinct. There was a pathos blended and infused into the melody, which caused a feeling of sadness to steal imperceptibly over me. While impressed with these feelings, Stephen commenced singing "Old Hundred," assisted by Mr. Frazer, whose voice was exceedingly rich and powerful. It was impressive beyond description, and I almost imagined myself in some vast cathedral, listening to the rich swelling notes of the organ.

Stephen died during the summer of 1857. He was buried on the hill south of the cave entrance. The small burial ground already contained the graves of three consumptive patients who had died in the cave. His widow Charlotte remarried. The last known reference to his son Thomas was five years before.

The cave management had lost one of its main attractions, but it was in their interests to minimize the loss. Mat, Nick, and the others continued to show the cave as before. Charles W. Wright's 1858 guidebook to Mammoth Cave stated: "Stephen, who had been a guide two years longer than Mat, died in July, 1857. Although a great deal has been said and written about him, from the fact that he was the favorite of the original proprietor, he was in no respect superior to either Mat or Nicholas, nor was his acquaintance with the cave more thorough or extensive."

No longer overshadowed by Stephen, Mat and Nick were now in demand as the most experienced guides. In 1863 Nick assisted an English explorer in making exciting new discoveries in the cave. Mat was with photographer Waldack in 1866 when he made the first successful cave photographs. Publicity resulting from these events, changing conditions at the cave, and the traumatic effects of the war years each tended to push Stephen's memory further to the background in the minds of the cave personnel. Alfred had died. During the 1860s Stephen's real accomplishments were all but forgotten. His name was seldom mentioned at the cave.

But Stephen's name still lived in Mammoth Cave literature.

Maryland physician W. Stump Forwood was writing a new book about Mammoth Cave. He wrote Franklin Gorin for more information, and Gorin answered:

> ...I placed a guide in the cave—the celebrated and great Stephen—and he aided in making the discoveries. He was the first person who ever crossed the Bottomless Pit. After Stephen crossed the Bottomless Pit, we discovered all that part of the cave now known beyond that point. Stephen was a self-educated man; he had a fine genius, a great fund of wit and humor, and some little knowledge of Latin and Greek, and much knowledge of geology; but his great talent was a perfect knowledge of man. There was not any Indian blood in Stephen's veins. I knew his reputed father, who was a white man. I owned Stephen's mother and brother, but not until after both children were born. Stephen was certainly a very extraordinary boy and man. He knew a gentleman or a lady as if by instinct. He learned whatever he wished, without trouble or labor; and it is said that a late professor of geology spoke highly of his knowledge in that department of science.

In Gorin's letter the story of the first crossing of Bottomless Pit had changed—after thirty years—in the retelling. Stevenson was omitted, so were the ladders. It was to change again in another thirty years.

Forwood quoted Gorin's letter in his new cave book, which appeared in 1870. The book was an instant success and replaced all previous publications as the standard reference book about the cave. Three printings were sold out within five years. The fourth edition was issued in 1875. The wide circulation of Forwood's book with its reference to Stephen created new interest in the name of Stephen Bishop. New Stephen stories appeared, some of them apocryphal, most of them dramatic. He became a folk hero to a new generation of cave personnel. His admirers were soon telling the visitors about daring deeds of Stephen; and the visitors were delighted. Appropriate stories were told when the visitors peered down into the depths of Bottomless Pit—

when they crossed the waters of Echo River—when they gazed with awe at the beauties of Cleaveland Avenue. The stories grew in the retelling until he became "the first guide and explorer of Mammoth Cave." The legend of Stephen Bishop, as we know it today, had its beginnings during the 1870s.

Pittsburgh millionaire James R. Mellon visited the cave for a week during November 1878. He was charmed by the stories he heard of Stephen Bishop. Charlotte, who was in charge of the hotel dining room, led him to Stephen's gravesite "which had only a cedar tree to mark it." He promised her that he would have

Grave marker lettered in Pittsburgh, Pennsylvania, in 1881, 24 years after Stephen's death in 1857. Note the wrong date. Photograph taken 1889 by C. G. Darnall. *Courtesy The Kentucky Library, Western Kentucky University, Bowling Green, Kentucky.*

a stone properly carved for Stephen's grave.

Three years after Mellon returned to Pittsburgh, he remembered his promise to Charlotte and arranged for the carving of the stone. In 1881 the stonemason, using a second-hand tombstone that a Civil War soldier's family had been unable to pay for, chiseled off the soldier's name and in its place carved the name of Stephen Bishop. The inscription on the finished stone read: "STEPHEN BISHOP, First Guide & Explorer of the Mammoth Cave, Died June 15, 1859 in his 37 year." The stone was shipped to the cave and erected at Stephen's grave. The error in the date of death detracted nothing from the legend now reinforced by a permanent record in stone.

The legend was further embellished during the 1890s. The story was told that "Stephen first crossed Bottomless Pit on a slender cedar sapling." It captured the imagination of the cave personnel and the public as well. Cave authors Horace C. Hovey and R. Ellsworth Call repeated it as history in 1897. Since then, the legend of Stephen Bishop has become a part of the cave's folklore.

The emerging image of the historical Stephen reveals him to have been a truly remarkable human being whose accomplishments and contributions were in the best traditions of Mammoth Cave.

The Caverns of Sonora

The Caverns of Sonora in Texas have been described as "the most beautiful caverns in the world." When the caverns were first discovered the location was kept secret: it was feared that hordes of visitors would destroy their fragile beauty. But James L. Papadakis felt that the caverns should be made available for more people than the handful of cavers who were able to negotiate the difficult ledge to reach the most beautiful formations. Carefully he began to develop the caverns, preserving the delicate features while at the same time making them accessible for the general public. Today, a tour of Sonora's expertly lighted and well-constructed trails can be experienced by everyone.

Jim Papadakis is a geologist, and he is also the developer of Crystal Ice Cave in Idaho. Both Sonora and Crystal Ice Cave have been designated as U.S. National Landmarks. Papadakis is a Fellow of the National Speleological Society; his outstanding cave photographs have been widely published.

The Caverns of Sonora

JAMES L. PAPADAKIS

It was Labor Day weekend, 1955. Jack Prince carefully felt his way along the rocky ledge of the dark, humid cave. He could hear the small noises of the other five men who were closely following his progress by the light of their headlamps. The six cavers acutely sensed the sixty-foot vertical drop below them, even though they could not see through the blackness which surrounded them.

Inching his way along the ledge Jack was determined to make it to the other side of the huge pit. To his right the cave wall sloped steeply upward, then rounded abruptly to a flat ceiling above his head. To his left the floor sloped downward at a forty-five-degree angle for three feet and then dropped steeply to the pit below and the jagged rocks within it.

Actually the ledge overhung the pit by several feet and Jack's awareness of this did nothing to minimize the tension. Much of the thin lime crust crumbled underfoot and handholds were next to impossible. The sixty-foot traverse to a wider and flatter part of the ledge must have seemed more like 120 at that moment.

Jack's trailbreaking was watched closely by the other members of the exploration team. Then each caver in turn crossed the pit—later to be known as Devil's Delight—without incident. As Jack suspected, a crawlway continued beyond the large pit room and the men pressed forward eagerly. Jack Allen, Peter Cobb, Allen Cain, Claude Head, and Daniel Sheffield were members of

---

The sturdy railings built by Papadakis and his group are still in use today. *Charlie and Jo Larson*.

the Dallas Speleological Society. All were eagerly anticipating discovery of something new and important.

The six men progressed alternately through low crawlways and larger walking passages for some five hundred feet. The nature of the cave seemed to be changing. Strange, amber-colored, wormlike crystals were seen on some of the walls and ceilings, and occasionally there were long, delicate sodastraw stalactites. The warm air became even more humid.

Jo Larson gazes at a group of stalagmites. The indirect lighting helps bring out the beauty of the speleothems. *Charlie Larson*.

Emerging from a narrow tunnel, the explorers stopped at the top of a slope that led to a spacious room. Dimly the combined lights of their carbide headlights lit the virgin scene in front of them. The floor, walls, and ceiling were completely covered with exquisite crystalline speleothems. It gave them the feeling of being enclosed within a giant crystal geode. The men could not proceed without damaging some of the delicate structures underfoot, and yet it was equally impossible to resist the temptation not to proceed. The adventurous cavers were lured deeper into the cavern; at every turn they beheld more unusual formations, all breathtakingly beautiful.

Finally, they were at the top of a vertical drop into what later was named the Lake Room. A decision would now have to be made whether to send someone back to the pit for rope or to temporarily abandon their efforts to continue. All of them were tired. They had been in the warm 72-degree cave since morning and had caved all the preceding day. Reluctantly they turned back, carefully recrossing the ledge by the pit.

The cave in which Jack and his friends made their great discovery that day would come to be described, perhaps justifiably so, as "the most beautiful cave in the world." It is located almost in the center of the vast Edwards plateau of central and southwestern Texas. The nearby town of Sonora lies 150 miles northwest of San Antonio. The most extensive cave systems in Texas are in the Cretaceous limestones of the Edwards plateau. The Balcones Escarpment, a major fault zone, forms the plateau's southern and southeastern boundaries. This cliff stretches for more than 200 miles. It rises 1,500 feet above the plateau near the Rio Grande, decreasing to less than 100 feet at its eastern edge. The movement of groundwater southward and eastward through the plateau has formed the numerous caves in the thick beds of soluble Cretaceous limestone. Much of this water surfaces as springs at the base of the Balcones Escarpment.

Although Jack Prince and his group deserve full credit for the discovery of "the pretty part," as it was later called, many people were and are involved in the story of this cave, which is now open to the public as the Caverns of Sonora. No one knows for sure when the single tiny entrance was discovered, or by whom. Local residents claim to have known of the cave and to have visited it thirty years before the 1955 trip. It is located about fourteen miles southwest of Sonora on the Mayfield ranch and thus was first known as Mayfield Cave.

In 1927, so one discovery story has it, a cowboy found a ten-inch hole and enlarged it to about eighteen inches, large enough to crawl through. I examined the entrance in 1957 and found that it had been chipped to a larger size. Also, the chipped surface was not fresh—it could have been done thirty years ago, or even earlier.

Bob and Bart Crisman of Abilene learned of Mayfield Cave and visited it in June of 1955. They were experienced cavers, possibly the first such to see Mayfield. They explored the part of the cave then known and came to the same pit that so many before them had seen. However, Bob and Bart knew that the tunnels they had spent hours exploring had carried great quantities of water at one time. It was not likely that everything ended at the pit. But were there any open passages beyond? On the far wall, near the ceiling, an opening was visible. Bob and Bart had been in the cave for hours, and Mayfield was not even one of their main objectives on this particular trip; at that time other nearby caves were considered far more interesting. They made no attempt to cross the pit, but they reported their observations to the members of the Dallas Speleological Society, who were also planning trips to the area.

After the historic discovery by the Dallas group in September 1955, the exploration and mapping of the Mayfield Cave was carried out with thoroughness and dedication. The cave extends about one-half mile from the entrance northeastward to the Helictite Room, with two additional main sections, the Hall of the White Giants and the Diamond Room Passage. The entire system is a complex network of passages on several different levels. About one and one-half miles of passage have been mapped and at least an equal amount is explored, but unmapped. Only a few very difficult leads remain unchecked.

By mid-1956 the cavers had essentially completed the exploration of a unique cave. It is true that other caves have sections resembling Mayfield, but no cave has yet been found containing such a proliferation of exotic speleothems. The Cretaceous limestone from which the cave was dissolved supplied the raw material—calcite—for its adornment. Calcite occurs in more forms—over 600—than any other mineral, and it probably would not have been too difficult to convince a visiting caver that all of them were represented here. Actually the principal forms of calcite represented include stalactites, stalagmites, flowstone, sodastraws, helictites, crystals, popcorn, and moonmilk. With many combinations possible, the result is a tremendous variety.

The names of some of the rooms and speleothems give one an idea of this variety. The rooms: Sodastraw, China Shop,

The Caverns of Sonora are considered by many people to be the most beautiful in the country. *Charlie and Jo Larson*.

Chandelier, Corinthian, Halo Lake, Christmas Tree, Valley of Ice, War Club, and Planetarium; the speleothems: Knotted Rope, Giant Candles, Dancing Serpents, Porcupines, Crystal Tree, Ice Cream Sundae, Elephant Tusks, Python's Den, Hanging Gardens, Angel Wings, Red-Topped Stalagmite, the Fishtails, Popcorn Trees, Moonmilk Falls, and, the most famous of all, the Butterfly. Color is usually delicate and ranges from pure white into shades of amber, tan, orange, brown, red-brown, red, and even tints of green. Most formations are translucent.

Under ultraviolet light most of the speleothems are fluorescent, turning a vivid blue. Even rarer is the phosphorescence, or afterglow, of the speleothems. This is most strikingly displayed when flash pictures are taken in a completely dark chamber after all eyes are adjusted to the dark. On signal, everyone closes his eyes and the flashbulb is set off. Immediately everybody opens his eyes again. The entire room glows with an eerie blue-green light. It quickly loses intensity, and in about three seconds the room is completely dark.

One of the things overlooked by many observers is that the entire surface of a speleothem, regardless of shape and size, may be wet, even though it is not being dripped on. This means that it is growing (or at least changing). This is only possible because the humidity stays at one hundred percent. The constant 72-degree temperature is no doubt an important factor as well.

The cave as we see it today is the result of a complex history of events. The original cavities were dissolved out of limestone by waters moving slowly from the northeast to southwest. This was controlled by the jointing and bedding planes in the horizontal limestone. The cave, then, is a tiny part of what was once a vast underground drainage system. Since the present water table is almost 200 feet below the deeper parts of the cave, this was a long time ago—probably even before the Pleistocene or ice age.

As the water table dropped, parts of the cave filled with air and speleothems developed from drip, flow, and evaporation. Then the water table rose again and began dissolving the newly deposited formations. Such fluctuations of the water table occurred several times. Finally the water table dropped so much that it no longer flowed through the cave, but left large standing pools. The great subaqueous coral-like deposits formed at this time in the pools. The cave has been drying out slowly ever since then as the water table continued to fall, draining the pools. As erosion proceeded in a canyon southwest of the cave, a tiny entrance was exposed. This further dried the air in the southwest section by evaporation.

To control traffic and prevent widespread damage to the delicate speleothems, an iron gate was installed over the en-

V. S. Watson is surrounded by splendor in the Helictite Room. The fragile helictites were deposited on existing stalactites and stalagmites when the room reflooded after the original speleothems had formed. *James L. Papadakis.*

trance soon after the discovery of the new section. In 1957 the Texas Regional Association of the National Speleological Society placed a register in the cave a short distance from the entrance. Everyone was to sign in and out. At the register a large sign read: "Take nothing but pictures! Leave nothing but footprints!"

I first learned of the cave in November 1956. At this time it was being called Secret Cave to help conceal its location. When an article appeared in the *Houston Chronicle* I contacted its author, Jimmy Walker. He couldn't reveal the location but referred me to the Crisman brothers.

Bart Crisman sent me the location; finally, on June 15, 1957, after picking up the key at the Mayfield ranch headquarters, Jack Burch, V. S. Watson, Ollie Testerman (all from Oklahoma), and I entered the Secret Cave for the first time. At the pit standard procedure now called for one caver to cross the Devil's Delight attached to a safety line. Upon reaching the other side he tied onto a sturdy stalagmite and the line was made tight. The next caver with carabiner on his safety line hooked onto the tensioned line, used it as a hand line, and walked across.

Like the others before us, we were overwhelmed by the beauty of the cave. Ours was a reconnaissance and photography trip. My primary interest was photography, and when we reached the "pretty part," even on that first trip, I felt compelled to photograph as much as possible. Many of the fantastic speleothems just were not going to survive the heavy visitation the cave was receiving. I was right. One of my pictures appears in "Exploring America Underground" by Charles E. Mohr in the June 1964, *National Geographic* Magazine. The caption reads, "Snake-dance helictites, since collapsed." It should have read "smashed by careless cavers."

Many other cavers were also concerned about the preservation of these extremely delicate speleothems. Foremost among them were Bob and Bart Crisman. At the 1957 Texas regional cavers' convention at Boerne, Bob showed their before-and-after series of color slides. It was very sobering to see pictures taken on the first trips and then see the identical scene on later trips. Many speleothems were ruined. Later? The cave had only been discovered two years before. This was not sobering, it was disaster.

It was August 3 before I returned to Secret Cave, this time with a small group of cavers I had organized in Houston. Four more trips followed in the fall of 1957. On these trips I realized that I had not perceived the extent of the damage on my first visit. Inexorably the area of human destruction was spreading. No one was supposed to wander off the established trails. Yet new trails were being started frequently through the delicate environment. Some major speleothems had been broken. Some

Jack Burch crosses the Devil's Delight in the Pit Room. Until 1955 the beautifully decorated rooms beyond this hazardous ledge remained undiscovered. *James L. Papadakis*.

cavers did not even bother to carry containers for their spent carbide but dumped it indiscriminately.

I never broke anything but tiny speleothems, but try as I might to be careful I managed to break a few on every trip. In this cave, when you are over six feet tall you feel like the proverbial bull in the china shop.

Stanley Mayfield, owner of the ranch, had never seen his own cave. He was a young, athletic man and I thought surely he could be persuaded into joining one of our expeditions. But I was never successful. Changing my approach one day I got his wife Elizabeth and their daughter Gerry excited about seeing the cave. Elizabeth implored her husband to go, but to no avail.

However, there were residents of the Mayfield ranch who visited the cave regularly. They had been exploring the "pretty part" a good many years before 1955. These were the little masked bandits themselves—the raccoons. On my way out of the cave one day I met a raccoon who was on his way in. We chanced upon a very awkward place to meet—the Devil's Delight. I was thirty feet behind one caver who was on the ledge. Evidently the raccoon had successfully hidden in some reentrant as my friend passed. Then, thinking the way was clear, he came out onto the ledge just as I emerged from the little tunnel. Hopelessly trapped, the raccoon quickly looked both ways, put one paw up to his chin as if thinking what to do, and then, without hesitation, went over the edge. I listened in vain for the "splat" of the small body on the rocks below. But the raccoon had negotiated the undercut and the vertical wall in darkness—such skill and tenacity!

The cave's permanent residents are mostly insects. Because of the unusual environment the chances of finding rare forms of life are good. Thomas C. Barr, Jr., described a new species of carabid beetle, *Rhadine babcocki,* in 1958.

Another itinerant visitor was Janie Taylor of Houston. She loved the cave and somehow could get around without ever breaking anything. Joining the 1957 expeditions, she was of great help to me with the photography. One would think the pictures I took of her in the cave were the only shots in existence—at least based upon published photographs. On several occasions I used Janie's hand for scale. Little did we realize that in three years hers would be the most publicized girl's hand in Texas!

Many Texas cavers were concerned about the extensive blasting near the cave for a natural-gas pipeline. I was in the cave in November 1957 when charges were being detonated about one-half mile from the cave, about the closest point reached. Fortunately, their concern was unwarranted, since the blasting was barely audible in the cave. It sounded like someone lightly tapping the cave wall with a hammer, a hundred or more feet away. The vibration was far too weak to break even the most

The famous Butterfly—and Janie Taylor's hands. *James L. Papadakis*.

fragile speleothem.

In August 1958, I was back to the cave on a three-day trip with Jack Burch, Ollie Testerman, Calvin Perryman, and Jim Schermerhorn. This was a photography trip. Jim snapped a picture of Calvin and me as we were photographing each other—all simultaneously! Our objective was to photograph the less-visited parts of the cave, where some of the most outstanding speleothems were to be found. We stayed underground so long that the high humidity softened the plastic coating on our flashbulbs. In the bell-shaped Dome Room they began going off like cannons. We were terrified, and ended that photo session promptly.

In 1959 I had the opportunity to bring the cave to the attention of the National Park Service and talk to one of their representatives about possible National Monument status. The location of the cave on private land presented some problems. Assuming the owner wanted to sell, it would still be costly for the Park Service to acquire. Personally I doubted that Mayfield would ever sell, but I didn't tell that to the government man. After listening to my description of the cave he concluded that any development for public use would be prohibitively costly; the National Park Service wasn't interested.

Then I discovered that not all the broken speleothems were still in the cave. Arriving unexpectedly in Dallas I found a good friend on his living room floor, itemizing his beautiful collection of speleothems—all from Mayfield Cave! One small specimen of a delicate honey color caught my eye. It was like a large transparent gumdrop completely studded with calcite crystals. It was a beauty. Since I showed an interest in it, he willingly let me borrow it in my efforts to acquire financing for the development of the cave for public visitation. Later, in Miami, a mineral dealer asked me how much I wanted for it, even though I had not given him the slightest hint it was for sale.

Most mineral collectors respect and protect unusual cave minerals. But it was almost obvious that if some unscrupulous rockhounds ever stumbled across the Butterfly deep in the cave they would not be able to sleep until they figured out a way to get it out of the cave in one piece.

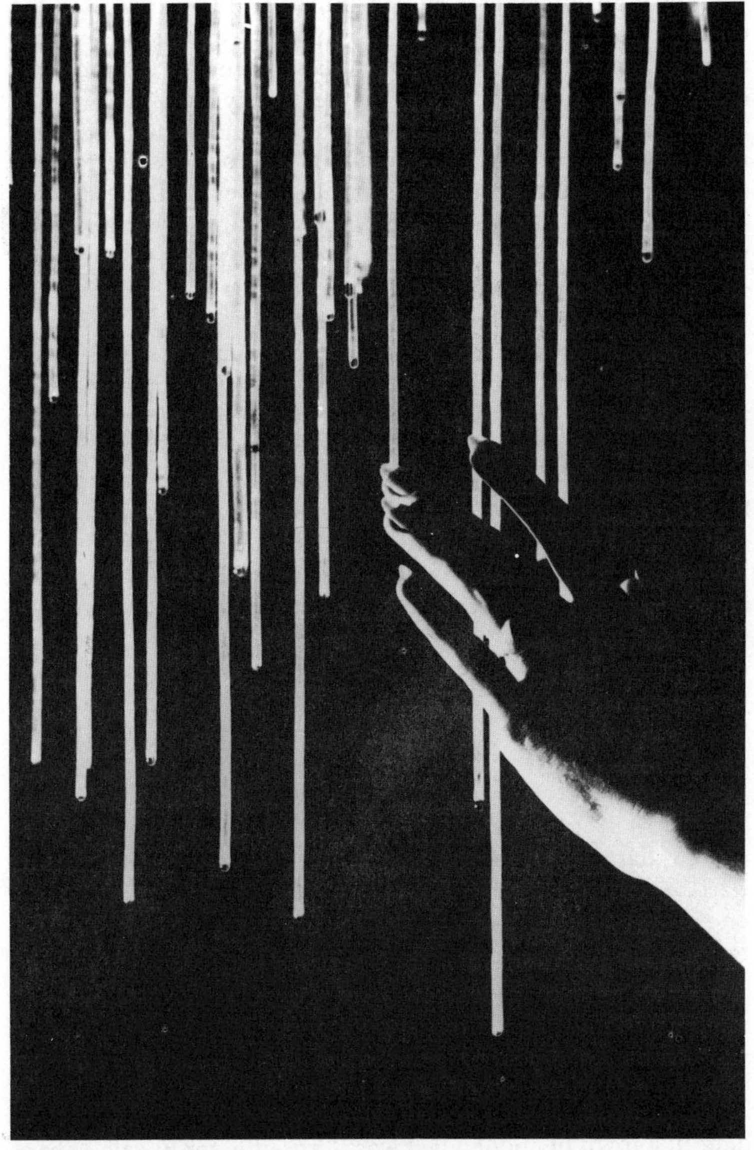

A group of sodastraw stalactites in the Soda Straw Room. Water seeps down and through the hollow inner tube and drips off the end. *James L. Papadakis*.

During September 1959, Jack Burch and I again visited the cave for three days and two nights. This trip was quite businesslike, as we laid out a feasible tourist route for a trail from the entrance through the Pit Room to the "pretty part." This developmental work could be done by a small crew within a few months. Jack proposed that we blast a trail out of solid rock on the west side of the pit across from the Devil's Delight. This would leave the historic route untouched but in full view of the public. I agreed completely. We ruled out any idea of tunneling directly into the "pretty part." We felt the public would be interested in seeing the discovery route followed in 1955.

Calvin Perryman first suggested the name "Caverns of Sonora." It was, and is, a beautiful name and most appropriate for the cave. It identified with the town which was only fourteen miles away. So, "Caverns of Sonora" it was.

The project was finally under way. I was negotiating with Stanley Mayfield on the lease. One major problem was an access road: Stanley did not want to take the route through the main ranch entrance. Since this would almost cut the nine-square mile ranch in half, I agreed that a better route should be found.

The second-best route left the main road a half mile past Mayfield's turnoff. It would be necessary to construct a road over a hill to the cave on the other side. Unfortunately the turnoff point was not on Stanley's land but belonged to a neighbor. And he flatly refused permission.

The third-best route left the main road near the boundary of Mayfield's property, climbed the limestone hill to the crest, and turned sharply to the left. The following spring Sonora contractor Carl Cahill blasted over this ridge and built a good gravel road for one and one-half miles to the cave. Interested in seeing the new business get off to a good start, he said I could pay him later.

While working on the lease with Mayfield I met H. V. "Buzzie" Stokes. I have never met anyone before or since like Buzzie. He was setting himself up in an office as new manager of the Sonora Chamber of Commerce. It was really an experiment because he was the first full-time manager the Chamber had ever had. In a town of about 3,000 was it possible for Buzzie to earn his salt? He did, and his public relations were a great help to us.

At the end of February the contract with Mayfield was signed. Work could commence after lambing in March when more help was available. Jack Burch from Springer, Oklahoma, and Mills Tandy from Ozona, Texas, joined the crew. We moved into an apartment until we could move a house onto the site. In March work on the cave began in earnest. The entrance was blasted open and the trail work was well on its way.

Since directional signs are so important to guide the public to an attraction, I set out to establish several sites on the main highway, U.S. 290, now Interstate 10. I discovered that most of the land belonged to the reluctant neighbors. As I looked out across the miles of seemingly endless rolling limestone hills I tried to think of a solution. We couldn't even get permission to put up a sign.

Luckily, the intersection was just inside the boundary of another neighbor of Mayfield's—Dock Simmons. When Clawson White and I first met with Dock he was against any signs on his land. But later Dock changed his mind, and he became one of the best friends a person could have.

Back at the cave, trail building was proceeding swiftly. We were at the Pit Room on the side across from the Devil's Delight. Now we faced a cliff of solid limestone and had to build a trail. Like a human fly, Jack Burch scaled the rock face and drove the blasting holes with the air drill. Again and again the Pit Room reverberated as we relentlessly carved a trail along the entire length of the cliff. The trail would then have to loop through a small room, which looked like a balcony on the sheer wall, and swing back over itself via a short bridge. A long cut would then bring the trail to the rooms beyond.

Once while blasting this cut we forgot to remove our light. Nobody felt like going all the way back and taking it down, so we decided to chance it and blast with the light still hanging. At that moment visitors from Sonora appeared. We told them to stand by because we were setting off a charge. The blast went off and the wildly gyrating light bulb disappeared in flying rock. After things settled down, we saw the bulb, still quivering but shining as brightly as ever. One of our guests asked incredulously, "Don't you take the light down when blasting?" "No," I said.

"We're getting so good at it we just blast around the bulb." The group left shaking their heads.

Now we had our pit crossing, but without a railing no tourist could be expected to walk on it. Stanley Mayfield came to the rescue. "I'll build a steel railing. I've always wanted to get one of those small portable welders anyway." For days the crackling electric welders sent weird shadows dancing across the Pit Room as Stanley and his neighbor, L. P. Bloodworth, put together a remarkable steel railing. Closely spaced rods between the main supports insured that even the smallest child could not slip through. Now I felt confident that an opening day could be established. I told Buzzie, who was now handling our publicity, that we would have the cave ready to open by July 16, 1960.

Conditions at the cave site were improving steadily. The housemovers finally had the house on its foundation. A 350-foot well was dug and a water system installed. Then in early May my mother and father arrived. My father, Louis (Louie) Papadakis, was looking forward to helping with the cave work. When my parents first arrived I was in town and Jack was working alone in the cave. Noting that my car was gone Dad thought we had all gone to town and left the cave lights on. So he threw the master switch. Snap! A quarter of a mile back in the cave Jack was in the dark without a flashlight. He had to feel his way out!

Dad would usually follow behind us and do the final touching up on the trail prior to the final coating. One day he returned from town with the largest pick he could find. Of the many different necessary tools, this was his favorite. One day Buzzie stopped by and wondered how we could build trails so fast. Louie revealed the secret: "I just put the boys in front of me and start swinging my pick. If they slow down, I start catching up to them. They have no choice but to build the trail fast!"

Meanwhile, in Sonora Buzzie was well under way with his promotional program for the grand opening. On one of his many visits to the cave he asked me to tell him all about the original discovery of the entrance. I mentioned the research that Mills Tandy had done and that quite frankly we didn't know who found it or when. As Buzzie wandered off, I heard him mutter something about it being bad for publicity. In the next issue of the weekly Sonora paper, *The Devil's River News,* I read Buzzie's solution to the problem: "The caverns were first found by a sheepherder around the turn of the century."

Undoubtedly, the apotheosis appeared in the March 1964, *Ford Times:* "The time was the turn of the century, and the place was Sonora, Texas, sixty-eight miles south of San Angelo, deep in the heart of Sutton County. An old Mexican sheepherder, guiding his animals over the scrubby hillsides, suddenly noticed steam drifting up from a small hole in the barren ground. Excited by his find..."

Bill Stephenson, founder of the National Speleological Society, visited the Caverns of Sonora on his way to the Society's annual convention in 1960, held at Carlsbad Cavern. Bill praised Sonora as the most beautiful known cave in the world.

July 16 was approaching fast. The trail through the most spectacular section was complete but the lighting was not. Mills Tandy had left for summer school. However, Calvin Perryman gave us some badly needed help. No compromise was made on the lighting. It had to be indirect with each bulb carefully positioned. Ivory zip cord to each light was carefully woven through the rough projections on the floor. It was almost invisible, particularly when we looked for it.

By now Buzzie was almost literally buzzing. With every press release the description of the Caverns of Sonora became more vivid. The Chamber of Commerce printed thousands of jumbo postcards with three scenes of the cave. Buzzie wrote in *The Devil's River News:* "If 100 people mailed 100 postcards each, we'd have 10,000 enthusiasts for the Caverns of Sonora. Let's start a mail campaign that will put Sonora and its Caverns on every map in the country. This is a big thing—let's do something about it."

Buzzie had everything planned to the last detail. Whenever anyone visited the cave, we had a real Texas barbecue. The air in the hollow where the cave's picnic area was located was permeated with the odor of barbecued meat. The biggest one of all was set for the grand opening on Saturday, July 23.

The first paying customers (after Joe VanderStucken, Sonora rancher who still holds ticket Number One) were ninety-two Boy

Scouts and leaders from Dothan, Alabama, who were on their way to a jamboree in Colorado. They were given the first guided tour at 10:00 P.M. Friday evening.

On Saturday, July 16, 1960, the Caverns of Sonora officially opened—and it rained. But the rains didn't seem to slow the crowds, as 2,000 people came for the tours the first weekend. Everything went smoothly, thanks to a lot of extra help in guiding from Mills Tandy and V. S. Watson, who were expert guides.

We were in high gear readying ourselves. Buzzie had the newspapers full of cave pictures. Some were pictures I had never seen before, and didn't even know existed. Buzzie was doing his job well.

In a moment of creative inspiration Dock Simmons decided he would like to play the organ in the cave on grand opening day. I didn't ask anyone how he played because I really didn't want to know how well or how poorly Dock played the organ. I had never even heard him perform. Our highway signs were on his land, and the choice was rather simple: either Dock played the organ and we had highway signs—or he didn't play and we didn't have any highway signs.

On grand opening day Dock's big electric organ was hauled 600 feet into the cave and set up in the Press Room. This was a large room just off the Pit Room. From the trail the room appears as a huge bowl with only the upper part visible. It is impossible to see anyone in the room from any part of the tour trail.

The night before, a large delegation from newspapers, radio, and TV stations attended an underground cocktail hour after taking a grand tour of the Caverns. As the evening progressed, the activity became a little frenzied. By the next morning, as the men struggled with the big electric organ, the alcohol-to-air ratio was pretty high. When the dedication ceremony was over and the crowds started surging through the cave, Dock was ready. Intoxicated by the grand opening ceremonies (and by the vapors from the previous night's libation), Dock played magnificently. As the groups filed back and forth through the Pit Room, the vapors were intertwined with the notes of "Also Sprach Zarathustra," and other selections, all emanating from that huge, mysterious spelean cavity. No one saw Dock, but everyone heard him. He was superb.

The people came, the tours were many—the dedication weekend was a success. Many thanks for this had to go to the superb execution of Buzzie's plans. Stanton Bundy, publisher of *The Devil's River News*, printed a special Caverns Edition. The masthead proclaimed, "Over 4,000 copies this week." The head-

Some of the railings are as much for the protection of the cave as they are for the safety of the tourists. *Charlie and Jo Larson*.

line screamed: "MORE THAN 2,000 PERSONS ATTEND CAVE DEDICATION." Mills Tandy's excellent article, "The Caverns of Sonora, Their History, Description, Exploration," was included in this edition.

Later in the summer, with the large crowds on weekends, I initiated a self-guided tour. Guides were stationed at strategic points along the trail and the visitor could move at his own pace and ask questions of the guides. It worked very well. I think we were the first cave to try this, and many others have followed our example.

In 1961 I sold the Caverns of Sonora to nine Sonora people who had formed a corporation with Stanley Mayfield at the head. Before I left Sonora I carefully explained what I had planned for the cave in the way of development. Of course I could make suggestions, but the fate of the Caverns was up to the others, and my role had ended.

Development continued. Stanley Mayfield was working underground on some trails, using the air hammer. He jarred a four-carat diamond out of his ring. The crew spent days sifting through the sands looking for it before they finally gave up. Probably it's still there today. So the Diamond Room Passage acquired its own real diamond! In June of 1962 finishing touches were put on a $60,000 visitor's center and the Caverns began another summer season.

In April 1963, the Texas House of Representatives gave the Caverns a special citation. A year later Bob Hope referred to the Caverns as "older than Bing Crosby." On December 1, 1965, the National Park Service included the Caverns in the National Register of Natural Landmarks.

Near disaster struck the Caverns early in 1966. In the vicinity of the Butterfly, many speleothems began turning black and a heavy fungus growth infested the area. Careful investigation led to the almost certain conclusion that sewage waters were entering the cave. Sludge pumped from septic tanks had been hauled away and dumped directly over the cave. This created a tough problem for the staff, but eventually it was solved and little or no permanent damage resulted.

Over the years local developments occurred which have

directly affected visitation of the Caverns. The road was paved all the way to the cave. Nearby Amistad Dam on the Rio Grande was completed. Interstate 10 was finished with an interchange at the Caverns road.

In the National Speleological Society's monthly publication, *The NSS News,* the Caverns of Sonora have been featured on the cover ten times—more frequently than any other cave. And NSS critics—whose taste in caves is certainly more discriminating than that of the general public—have heaped praises on Sonora. Bill Stephenson of the NSS summed it up: "This is the most indescribably beautiful cavern in the world. Its beauty cannot be exaggerated... even by Texans."

Springs and Sewage

Springs, sinking streams, and caves naturally go together. In some cave areas there may be no surface streams at all—the drainage is all underground. Such features, along with sinkholes and other phenomena, often underlie the cave areas of the world.

Most springs, as they flow out of the ground, look clear and pure. The water must be good to drink—it's filtered by the rocks. Right?

Wrong! That clear, cool water may be highly polluted. Tom Aley, Director of the Ozark Underground Laboratory, spent several years studying Missouri's underground drainage. The water comes from someplace. He found that what goes into the ground must later come up, and that sewage dumped into the ground can travel scores of miles to reappear elsewhere at a spring—or even at the turn of a faucet.

Springs and Sewage

TOM ALEY

The Ozarks cover parts of Missouri, Arkansas, and a little bit of Oklahoma. Many people who drive through the area think of it as rather flat but with unexplainably crooked roads. Actually the Ozarks is a land of valleys with flat-topped ridges and plateaus. The roads, for the most part, follow the meandering ridge tops which are dissected or cut into by countless streams. The landscape is thus one of rolling hills and winding ridge tops. The traveler in a hurry is apt to miss the valleys and hollows which dissect the land. He will also miss the myriad caves and springs in these valleys which are integral parts of the Ozarks.

Some of the largest springs in the United States flow from the dolomite and limestone bedrock of the Ozarks. The largest of these, Big Spring, near Van Buren, Missouri, has a mean flow of 428 cubic feet per second—enough water to almost fill an average 10-by-12-foot room every two seconds. Water flow from all springs in Missouri is over a billion gallons per day.

It is axiomatic that whatever goes up comes down. But the opposite of that axiom is also true when one considers water

---

Water from this sewage lagoon at Mountain View, Missouri, disappears a few feet after entering the channel of Jam Up Creek. It flows underground to appear on the surface at Big Spring, 38 miles away. *Tom Aley*.

movement in limestone cave terrains. In these places, whatever goes down eventually comes up. This is true not only for the water, but for whatever may be in it. Contaminants or pollutants introduced underground will ultimately reappear on the surface through springs or wells.

I have talked with people in the Ozarks who believe that the water which flows from the springs must have originated in the Great Lakes or Rocky Mountains. How else, they ask, could one get the tremendous flows of water which are observed at the springs? They fail to consider, or perhaps fail to see, the almost perennially dry valleys and sinkholes of the region which channel tremendous quantities of water into the subsurface. The source of water for the springs comes from the local area, not from far-distant lakes and mountains. The source of spring flow is rainfall, which occurs on the surface and flows or sinks into the ground.

The large springs of the Ozarks provide subsurface drainage for extensive areas. In much of the Ozarks, more water moves underground through the spring systems than flows on the surface. Many surface streams with drainage basins as large as thirty to forty square miles are normally dry, and flow for less than a day after a major storm. A classic example of this is found on Hurricane Creek, a 113-square-mile surface drainage basin tributary to the Eleven Point River south of Winona, Missouri. I have done extensive hydrologic studies on Hurricane Creek, and have found that only about fifteen percent of the total annual runoff from this basin flows down the surface channel and enters the Eleven Point River. The remaining eighty-five percent of the annual runoff flows underground and discharges at springs outside the Hurricane Creek basin. Subsurface waters from the Hurricane Creek basin flow to both Graveyard Spring (on the Eleven Point River) and to Big Spring (on the Current River). Between sixty and seventy percent of the annual runoff from Hurricane Creek ultimately discharges from Big Spring.

In the region of Shannon, Carter, and Oregon counties, Missouri, the amount of area drained by a large spring can be approximated by assuming that there will be one square mile of drainage area (or recharge area) for each cubic foot per second of mean annual spring flow. Big Spring, as an example, has a mean annual flow of 428 cubic feet per second. Using the approximation, the recharge area for this spring is about 428 square miles. Greer Spring, the second largest spring in Missouri, has a mean annual flow of about 328 cubic feet per second, so its recharge area must be about 328 square miles.

When we consider the springs of the Ozarks, not only are we dealing with gigantic recharge areas, but we are also dealing with tremendous volumes of water. One cubic foot per second for a year equals 236 million gallons of water; Big Spring discharges about 100 billion gallons of water a year.

To determine the location of recharge areas of the large springs in the Ozarks, I began an extensive groundwater tracing program in 1968. I was then employed as a hydrologist by the United States Forest Service, and directed the Hurricane Creek Barometer Watershed Project. The purpose of the Hurricane Creek study was to relate use of the land to the quality and quantity of water discharging from the springs of the Ozarks.

We had installed precipitation gauges and stream flow measuring stations on the basin in 1966, and it was obvious that much of the water yield from Hurricane Creek was discharging from the large springs of the area. Our problem was to determine which springs drained what areas, and then to determine what factors affected the quantity and quality of spring flow.

A number of different materials have been used to trace groundwater movement. In the late 1800s painted live ducks were used in French cave systems. This technique must have been traumatic for the ducks, and the mortality rate of ducks unable to swim from a cave system through the solution channels to a spring was undoubtedly high. In Missouri, bales of hay, wheat chaff, and corncobs have been used by local people in attempts to trace water from sinkholes and disappearing streams to distant springs. Revenue agents in Tennessee poured 2,000 gallons of illegal whiskey into a sinkhole and inadvertently learned that liquids from this point flowed to a spring supplying water for a local high school. Fortunately, there are better tracing agents than ducks, hay, wheat chaff, corncobs, or moonshine.

On the Hurricane Creek basin I began working with fluorescein dye, a brilliant fluorescent green organic dye which has long been used in water pollution and water tracing work. In the first successful tracing I injected ten pounds of the dye into water disappearing at the Blowing Spring Estavella. An estavella is a stream which can either channel water from the surface into the subsurface at low flow conditions, or function as a spring at high flow conditions. At the time of the dye injection, water was disappearing into the estavella. When you put your ear to the ground, you could hear the water roaring underground. The Blowing Spring Estavella is in the normally dry channel of Hurricane Creek near the abandoned community of New Liberty.

Screen-wire packets filled with chemically active charcoal were placed at a number of springs in both the Current and Eleven Point River basins. The activated charcoal is capable of adsorbing fluorescein dye, even in concentrations as small as a few parts per billion. The system works because large quantities of water are passed through the charcoal packets.

Packets were collected once a week from all of the springs. It was with rather high hopes that I collected the packets the first week and began the analysis. A chemical is poured over the charcoal which replaces any adsorbed fluorescein, which, if present, can be seen hovering above the charcoal in the solution as a distinctive green dye.

In the first week the results were negative; no fluorescein was detected at any of the springs. As far as I could tell from the literature, nobody had tried to trace such a long distance to a major spring with so little dye. Equations in the technical literature indicated that I should have used several hundred pounds of dye, and I had used only ten pounds. My budget could stand ten pounds of dye, but more than that was out of the question.

Waiting for dye to reappear at a spring is a matter of wondering when, where, and if. I wondered those things for another week, then collected samples again and took them to the lab. The charcoal from each packet was put in a jar, and the chemical was poured in. Results were negative until I got to Big Spring. That sample immediately turned a brilliant, distinctive fluorescein green. I had just proved that water from Hurricane Creek flowed under a major topographic divide between two surface river systems and reappeared at Big Spring, seventeen miles from the Blowing Spring Estavella.

Since the first successful tracing of subsurface water to Big Spring, we have had several dozen additional successful traces in the Ozarks. Everett Chaney of Birch Tree, Missouri, and I have spent hundreds of hours in groundwater tracing work. Much of the tracing has been to define the limits of the recharge areas for the large springs in an area of about 1,500 square miles lying between the Current and North Fork rivers in Missouri. The work has been remarkably successful.

We have learned that Big Spring draws the majority of its flow from the Eleven Point River basin. Our longest subsurface tracing to Big Spring was from the channel of the Middle Fork of the Eleven Point River near Fanchon, Missouri. Water from this point leaves the Eleven Point basin and discharges at Big Spring, 39½ miles straight-line distance away. We have also learned that water flowing from the sewage lagoons at Mountain View, Missouri, sinks in the channel of Jam Up Creek and reappears at Big Spring, about 38 mles away. Water from a laundromat in Winona, 19 miles straight-line distance away, also reappears at Big Spring.

Springs in the Ozarks are important cultural features. Early settlers relied on the springs for drinking water, and many of the larger springs have been used to power gristmills and occasionally sawmills. Even today, the springs are important to the people of the Ozarks. It is frequently the local people, not the out-of-state tourists, who visit the springs in the Ozarks on weekends.

For many Ozarkers, a visit to a spring is almost a ritual. Everyone drinks the water and comments on its clarity, coldness, and purity. Clear and cold it is, but purity is a different matter. My concern with purity is not the academic point that the water flowing from limestone springs contains dissolved rock and other mineral compounds. My concern is with bacteria and viruses which may endanger health, and with nutrients and chemicals which may alter the very nature of the springs and rivers of the area.

Many people who drink water from these springs assume

that, since it has been underground, it has been filtered or otherwise purified. Many people in limestone areas think they live above an infinite filter which will remove any contaminants from the waters that go underground. Such is not the case; rather than living on a filter, those of us who live in limestone areas live upon a highly developed network of caves and solution openings. When water flows through these caves and solution openings, it is not filtered, but merely transported to the springs.

The question of whether effective natural filtration occurs in limestone groundwater systems intrigued me, so in 1971 I began tracing groundwater with stained *Lycopodium* spores. The common name for the genus *Lycopodium* is club moss; they are simple plants found at high latitudes. The spores are tiny, very hard, and nearly spherical in shape. The average size of the spores is about 33 microns; it would thus take 850 spores laid side by side to span an inch. Tiny as they are, the spores are still ten to fifteen times larger than most bacteria which cause disease in man. The spores are about 300 times larger than most viruses, and are 15,000 times larger than the virus of infectious hepatitis.

I reasoned that *Lycopodium* spores, being so much larger than pathogenic or disease-causing bacteria and viruses, would be an excellent agent for determining if effective natural filtration occurred in groundwater systems. In crude laboratory testing, I found that a foot of sand removed essentially all *Lycopodium* spores from water. Admittedly, bacteria and viruses may behave somewhat differently than the spores; for example, bacteria and viruses may tend to be adsorbed on charged particles while the spores are not adsorbed. Still, if the groundwater system cannot mechanically filter out something as relatively large as *Lycopodium* spores, I would not want to trust it to do an adequate job with bacteria and viruses.

Beginning work with *Lycopodium* spores caused one ridiculous problem after another. First I ordered fifty pounds of the spores and caused quite a stir at the biological supply house. The spores are typically used as a dry lubricant on prophylactics or, in small quantities, for classroom demonstrations of powder explosions. The biological supply house was not accustomed to hydrologists ordering fifty pounds of the spores, but finally filled the order.

When the spores arrived, I added some to a bottle of water, expecting to see them go into suspension. They floated, which was emphatically not what I planned. I shook the bottle; they still floated. I added detergent to serve as a wetting agent; they still floated. I waited overnight; they still floated. People kept coming by my lab and chuckling, but I didn't see anything funny. We finally decided to cook the spores in a big washtub over a portable stove in the parking lot. If all else fails, cook it! It worked, and from there on staining the spores with bacteriological stains and treating them with preservatives presented few problems.

Next we had some nets made of calibrated nylon netting which had openings smaller than the spores. These nets were placed inside pieces of steel culvert which were screened to keep out logs and crayfish. The *Lycopodium* samplers were then placed in springs where they would filter water. We would periodically collect everything caught in the nets and take it to the laboratory. After some additional filtration and treatment, we would examine the sample under a microscope to see if any of our stained *Lycopodium* spores, injected at some distant point, had reappeared at the spring.

Tracing with *Lycopodium* spores was far more successful than I had initially expected. We mixed the spores with water and injected them in a dump in a deep sinkhole. The dump was filled with trash, garbage, bones from dead animals, and even sludge from septic tank cleanings. A few days after the injection, we recovered the spores five and one-half miles away at a spring where several thousand people drink water every year. Remember, the spores are much larger than disease-causing bacteria and viruses. Imagine what people were drinking.

We also used the spores to trace water from disappearing surface streams to distant springs. To demonstrate the inability of spring systems to provide effective natural filtration, we used the spores to trace water from the channel of the Middle Fork of the Eleven Point River to Big Spring, a straight-line distance of thirty-nine and one-half miles.

To me, the most frightening tracing of all was only a few hundred feet long. Three days after flushing spores down a toilet we recovered them from a state-approved well which supplied water for a large family. This tracing was done because periodic

A sinkhole dump near Alton, Missouri. The sink is usually dry but during storm periods it partially fills with water, as shown here. Once the storm stops, the sinkhole drains and the contaminated water reappears 15½ miles away at Morgan Spring on the Eleven Point River. *Tom Aley.*

bacteriologic sampling for coliform (intestinal) bacteria indicated that there was an apparent problem with the well. The tracing showed the importance of such periodic sampling, and also highlighted the problems of sewage disposal in limestone areas.

Obviously, there are processes other than mechanical filtration that are important in the "natural purification" of water. On the surface, ultraviolet light from the sun is important in killing pathogens. Underground, there is no sun, so this process does not function in groundwater systems.

Pathogens, and other contaminants as well, can be adsorbed on charged particles such as clays. This adsorption of contaminants is similar to the adsorption of fluorescein dye on activated charcoal. The conduits which transport water from localized points of groundwater recharge to the spring systems provide little adsorption for contaminants. If the conduits did provide good adsorption, a dye such as fluorescein (which has a high sorption tendency) would be a poor groundwater tracing agent as it would be lost to adsorption. But fluorescein dye is an excellent water tracing agent in the Ozarks, in large measure because chemical adsorption is not an effective natural cleansing agent in the groundwater systems feeding the springs of the Ozark limestones.

Bacteria and viruses are live organisms, and everything alive ultimately dies. If the travel time for the underground water is sufficiently long, then the pathogens in the water will all die before the water again reaches the surface through a spring or well. Unfortunately, there are numerous factors which determine the life span of pathogens; there is no good way of telling how long is sufficiently long to insure the death of all pathogens. About the best we can do is say that the longer the water is underground, the greater is the chance that all pathogens will be dead before the water resurfaces.

Springs in the Ozarks typically respond very rapidly to surface precipitation. As an example, Big Spring typically reaches peak flow on either the same day or the day following the time the Current River (into which the spring flows) reaches peak flow. This occurs because water travel through the spring system is very rapid. In our groundwater tracing work we have almost always encountered very rapid groundwater movement. For example, water disappearing at the laundromat in Winona reappears about nine days later at Big Spring, which is over 19 miles straight-line distance away. It takes about twelve days for water to travel the thirty-eight miles from the Mountain View sewage lagoons to Big Spring. No one would call these travel rates slow groundwater movement.

Travel times of nine to twelve days will result in some, and perhaps substantial, pathogen die-off. But the travel times are not sufficiently long to insure that all, or nearly all, of the pathogens will die before the waters reappear on the surface.

Bacterial and viral contamination of groundwater can be serious in limestone areas, but there are other forms of contamination which are also important. Plant nutrients and nitrogen compounds are of great importance as potential contaminants in the Ozarks. The springs and rivers of the Ozarks are nationally known for their great clarity; typically you can see the bottom of the stream through fifteen or twenty feet of water. The addition of minute quantities of nitrogen can destroy this clarity. Nitrogen causes growth of substantially greater quantities of plant material in the water than occurs under natural conditions. This growth of plant material reduces the water clarity and substantially changes the nature of springs and rivers. A permanent increase in nitrogen of one part per million could destroy the uniqueness of the Ozark waterways.

Nitrogen compounds come largely from agricultural activities and from sewage and animal wastes. If these compounds enter a spring system, they will ultimately discharge at the spring. The spring systems serve as a conduit, conducting the nitrogen compounds from the land where they could be used, and where they should be retained, to the springs and the rivers where they can have highly detrimental effects.

There are many toxic chemicals, and many of these can also be transported through limestone groundwater systems. If there is insufficient adsorption to remove fluorescein dye from water moving through limestone spring systems, can we realistically anticipate effective adsorption of pesticides? I think not.

A classic case of chemical groundwater pollution occurred at Big Spring in the 1920s. At that time an iron company had a large ore-processing operation at Midco on Pike Creek, about ten miles from Big Spring. One of the waste products from the operation was an alcohol, which was dumped into the normally dry channel of Pike Creek. Pike Creek is usually dry because most of the water yield from the basin moves underground and discharges at Big Spring. Predictably, the alcohol from Pike Creek soon began to appear at Big Spring.

Big Spring joins the Current River about 35 miles upstream of the city of Doniphan. The municipal water supply for Doniphan was the Current River, and the alcohol became so noticeable in the city water that legal action was begun by the city against the company. The problem ended when the company went out of business. Today Midco is a ghost town with empty foundations and a towering brick smokestack, and Big Spring is no longer polluted with alcohol.

Study projects, and particularly the U.S. Forest Service study on Hurricane Creek, have demonstrated the sensitivity of limestone groundwater systems to contamination. Groundwater contamination can of course occur in any type of terrain, yet the hazards in cave areas are greater than in almost any other landscape. In what other terrain can you take your trash to a dump, and then a week later drink (if you are so foolish) of its effluent at a spring several miles away? Where else can you do your laundry and then, nine days later, see some of the same water emerge from a spring nineteen miles away?

Protection of limestone spring systems requires knowledge, diligence, and careful use of the land. We are making progress on this in the Missouri Ozarks, but there is much that remains to be done. The Ozarks is a landscape where the altered axiom "what goes down must come up" applies with a vengeance. Until we can adequately take care of whatever goes down, drinking spring water will be an Ozark variety of Russian roulette.

Carlsbad, The Giant

Carlsbad Cavern has a special fascination for me. One of my most interesting summers was spent there as a seasonal ranger, giving me a perspective quite different than that of a tourist or caver. Visitors had thousands of questions, and rangers were—and still are, I hope—expected to have the answers. Most questions were reasonable, but some were the classics heard by cave guides everywhere: "How many miles of unexplored cave are there?" "Is this cave all underground?" "What's it like down here at night?"

Carlsbad is unique. It is a large cave, but not the largest. It is a deep cave, but not the deepest. Other caves have larger formations, or more bats. What does make Carlsbad unique is its sheer volume and spaciousness. Large corridors and cathedral-size rooms follow one after another in pristine grandeur, culminating in the world's largest underground chamber, named—simply and appropriately—the Big Room.

The National Park Service has spent much time, money, and effort to prepare Carlsbad Cavern for the public and yet preserve the cave. Today, in perfect safety on a self-guiding tour, you can visit this natural wonder, and experience the grandeur of this underground wilderness for yourself.

TEMPLE OF THE SUN

Carlsbad, The Giant

BRUCE SLOANE

Much has been written about Carlsbad Cavern. And deservedly so. Since it came under the conrol and protection of the National Park Service in 1923, more than twenty million visitors have viewed its vast halls and formations beneath the southern New Mexico desert. Today over 800,000 people tour the cavern each year. Although there are many caves whose total length exceeds Carlsbad's nineteen miles of mapped passages, none can match it for the sheer size and magnitude of its rooms, and the grandeur and magnificence of its formations. The Big Room, probably the world's largest natural underground chamber, covers an area of more than twelve acres, with the roof arching in places 255 feet above the floor. Speleothems of all kinds decorate the spacious chambers, in places covering the walls, ceiling, and floor with formations of all descriptions—stalactites beyond count, stalagmites towering up from the floor half a hundred feet, festoons of rock curtains suspended in space, all in infinite variety.

The story of Carlsbad begins with its geology. The limestone in which the cavern is formed was laid down in a gigantic reef during Permian times, about 250 million years ago. At that time

---

The Temple of the Sun in the Big Room. *National Park Service.*

an inland sea covered a large portion of what is now New Mexico and Texas. In the shallow water toward the shore of this sea a barrier reef was built by lime-secreting organisms. The reef grew upward and seaward mostly on talus or blocks of rock broken from the seaward face. Slow sinking of the reef allowed virtually continuous growth until it had formed a rim around the Permian sea, similar to the barrier reefs off Australia today, which fringe much of that continent. Finally channels to the open sea were cut off, killing the reef. By then it had advanced seaward up to 20 miles, and was more than 2,000 feet thick.

Other sediments were deposited on top of the reef limestone. The weight and pressure caused a system of cracks or joints to develop in the rocks. Two sets of these joints were formed. One developed parallel to the trend of the reef, the other at near right angles to the first. These cracks or joints determined where the cavern would form. Groundwater then began the long process of dissolving the limestone, at first only small crevices along the joints. As time went on and solution continued, the passages enlarged, both by collapse and further dissolving. The near right-angle alignment of the joints created the arrangement of the main passages.

Faulting later lifted and tilted the entire block of reef limestone to form the present Guadalupe Mountains and Carlsbad Cavern. Erosion of surrounding rocks formed the long ridge. As uplift continued, the cavern slowly drained and filled with air. The large rooms with rock rubble floors owe their great size to the collapse of honeycombed blocks of solution-ridden limestone which fell to the floor when the supportive water drained away. Today about 40 miles of reef are exposed at the surface, comprising much of what is now Carlsbad Caverns and Guadalupe Mountains National Parks, and Lincoln National Forest with over three hundred known caves. Two other segments of the reef are exposed to the south in the Apache and Glass mountains of Texas.

As the cave filled with air, deposition of speleothems began through dripping water saturated with dissolved limestone. Presumably the huge stalagmites and other speleothems formed largely during the wetter periods of the Pleistocene or ice age

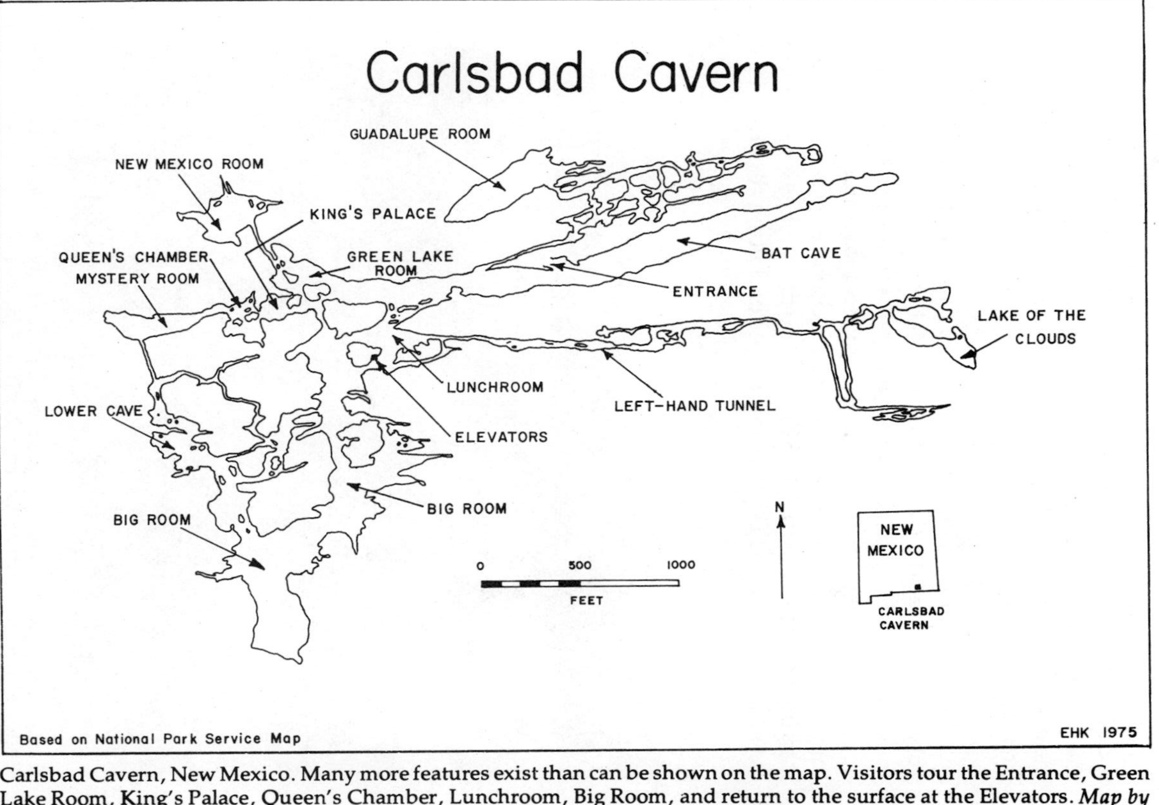

Carlsbad Cavern, New Mexico. Many more features exist than can be shown on the map. Visitors tour the Entrance, Green Lake Room, King's Palace, Queen's Chamber, Lunchroom, Big Room, and return to the surface at the Elevators. *Map by Ernst H. Kastning.*

when glaciation was taking place farther north. The cavern today is relatively dry with little dripping water, and, although there are some standing pools, few formations are growing or forming today. Alternating arid and wet climates caused fluctuations in the water table with some reflooding of parts of the cave. But for thousands of years, the cavern has existed essentially as when first discovered.

The first humans to see Carlsbad's yawning entrance probably were the Basket Maker Indians. But there is no evidence that they, or the later Apaches, ever entered the cavern. Although Spanish explorers in the 1500s passed close by, most probably they were not aware of the cavern's existence. Settlers began moving into the area in the 1870s, and no doubt they knew of the cavern. Stories of the first explorers are conflicting, but in 1903 a

The Guadalupe Mountains, an ancient reef, form a forty-mile-long ridge to the south and west of Carlsbad Caverns. *National Park Service.*

mineral claim was filed to remove the huge quantities of bat guano, which was used for fertilizer in the citrus groves of southern California. The guano was found in a passage extending about a half mile from the entrance, where millions of bats roosted during the day. However, the mining operation was never more than marginally profitable and was discontinued in 1941.

Most of the miners showed little interest in the cavern, except for the guano. But Jim White, a local cowboy who worked for a mining company, began to explore the farther reaches. He told others about his findings, built rough trails, and guided all who wished to visit. Early visitors were lowered into the cavern in guano buckets. Citizens in the nearby town of Carlsbad were skeptical at first, but became promoters of the cavern as a scenic attraction as its wonders became known. Although some pictures were taken as early as 1907, they were not widely distributed.

Today's visitor descends into the pit-like entrance on graded, paved trails. *National Park Service.*

Stories of the cavern reached the General Land Office, and a government employee visited it in 1923, despite official doubts that such a cavern could exist in the middle of the desert. But a month's study resulted in the recommendation that the cavern be designated a national monument. Dr. Willis T. Lee, a geologist with the United States Geological Survey, concurred with that recommendation. On October 25, 1923, Carlsbad Cave National Monument, consisting of 720 acres, was established by proclamation of President Calvin Coolidge. This was the first cavern to become a part of the national park system, and set the precedent for the later establishment of national park status for Mammoth, Jewel, Wind, and other caves.

But much work needed to be done before the cavern could be opened to the public. Dr. Lee, under the sponsorship of the National Geographic Society, spent six months exploring, mapping, laying out trails, and photographing the new national monument. His survey of the cavern and discovery of other significant caves nearby led him to recommend that additional acreage be added to the monument. In 1930, by an act of Congress, Carlsbad Caverns National Park was established. In later years the boundaries were extended, until at present the park contains over 46,000 acres.

Few present-day visitors are aware of the tremendous efforts of the National Park Service to make their trip as pleasant and interesting as possible. The guano bucket was still used to enter and leave when the cavern became a national monument, and visitors had to make their way on rough trails, with only dim lanterns to pierce the darkness of the huge galleries. In 1924, the first full year of its operation as a national monument, 1,876 visitors toured the underground wonder. Each year their numbers increased and improvements were needed in a hurry. A stairway into the natural entrance was built by the Carlsbad Chamber of Commerce in 1925, ending the rough descent by bucket. Dirt trails and wooden stairways were installed. Indirect lighting, which soon replaced the lantern, helped reveal the awesome beauty of the cavern. An underground lunchroom near the Big Room provided rest and needed sustenance.

Still, visitors to the cavern faced an arduous trip. The first

The immensity of the Big Room stretches off into the distance. *National Park Service*.

mile and three-fourths required a descent of more than 800 feet, mostly over huge blocks of rocks, followed by an eighty-foot climb before reaching the Underground Lunchroom. But the newly improved trails helped ease the trip. This was followed by a near-level walk of over a mile around the Big Room before returning to the Lunchroom. To return to the entrance, it was necessary to climb back up those 800 feet, retracing the first mile and three-fourths. In 1931 an elevator was installed next to the Lunchroom, making the long walk in and out optional. Visitors could now enter by elevator, make the relatively easy but spectacular trip around the Big Room, and then return to the surface by elevator. Today, with paved trails and other improvements, even people in wheelchairs can tour much of the Big Room.

During the early years, tours began as soon as a group of visitors had arrived. As attendance increased, tours had to be scheduled at specific times, and several tours a day were made. Groups walking out of the natural entrance could only pass entering parties at certain places along the frequently narrow trail. It was necessary to limit the size of the parties when the elevators' capacity was reached. Although new elevators were added, making a present total of four, a strict timetable of tours was necessary. Occasionally, because of large parties, some of the stops and talks on the trips were eliminated or shortened. Walkout tours were eliminated owing to the difficulties of two-way travel on narrow trails, and all visitors left the cavern by elevator. The Underground Lunchroom and the adjoining elevators became critical locations. On busy summer weekends, tours had to be run with railroad-like timing. Huge parties, sometimes as many as 1,800 people, made this even more difficult, since one walk-in group barely had time to eat in the Lunchroom prior to the trip around the Big Room before another one arrived; at the same time, a third party was arriving at the elevators to leave the cavern after completing the Big Room tour. And there were associated problems which had to dovetail with so many multiple parties: cleanup, maintenance, and construction all had to be scheduled at night; sewage and waste from the cave (which was pumped up the elevator shafts) and from surface operations had to be disposed of properly.

The Rock of Ages is one of the largest stalagmites in the caverns. *National Park Service*.

In the summer of 1968, scheduled guided tours of the Big Room were eliminated. Visitors were permitted to enter the Big Room as they arrived, proceed along the trail at their own pace, and exit when they wished. Rangers were stationed at places along the trail to answer questions, and booklets and interpretive signs were available. Most visitors gave a very favorable response to this system, and it was continued in the Big Room during the next several summers. Park officials felt that the quality of visitor enjoyment increased with this free-flow method and hoped to extend it to the entire cavern year round. This was accomplished in 1972, when more than 850,000 people took the semi-guided tour.

To further increase visitor enjoyment and understanding of the cavern, an electronic interpretive system was installed in 1974, similar to those used in many museums. Visitors now receive recorded narrations about cavern features automatically through specially designed portable listening devices. More than 40 messages are heard as one progresses through the caverns, with identical Spanish-language messages and a special children's version, each on a separate channel. Park rangers are still stationed throughout the cave to answer questions, and to protect the cave.

After receiving his portable listening device, today's visitor is greeted at the cave entrance by a ranger. He then descends over a winding but graded paved trail, to the level of the Bat Cave. The trail then follows the main corridor for nearly a mile, ever downward, over piles of rubble and huge rocks, with the ceiling rising as much as 200 feet above the trail. A tour of the Green Lake Room, King's Palace, Queen's Chamber, and other scenic areas 830 feet below the entrance is followed by a short climb to the Lunchroom. After a pause here, the trail encircles the Big Room, a T-shaped chamber whose main arm is more than 1,800 feet long with a cross-bar extending 1,100 feet. Massive stalagmites tower fifty to sixty feet high, with equally magnificent stalactites and other speleothems. A gaping pit reveals a tantalizing view of the undeveloped Lower Cave. The trail returns to the Lunchroom, where the visitor returns to the surface by high-speed elevator. This complete tour is a walk of about three miles. For those entering by elevator for the Big Room tour only, the trip is about one and a quarter miles. The total known length of the cave is 19.2 miles, including the Guadalupe Room, recently discovered by the Cave Research Foundation, although final mapping is not yet complete. For many years, before accurate surveys, the reported length was 23 miles. There are many caves that are considerably longer, but none has Carlsbad's combination of cathedral-size rooms, gigantic corridors, and stupendous formations. The deepest point in the cave—1,024 feet below the entrance—is near the end of the Left Hand Tunnel. Although visitors do not see this and other undeveloped areas they do see the huge rooms, galleries, and speleothems that have made Carlsbad unique.

For many visitors the bat flights are as interesting as the cavern. Over ten species of bats have been identified in the cavern, but most of them are Mexican freetailed bats that make their summer home in the Bat Cave. Each evening from April through October several thousand bats fly from the entrance to begin their nightly foraging for insects and water. Once numbering in the millions, the bats reportedly led to the discovery of the cavern as early settlers saw what they thought was smoke rising from the hills. The best flights usually can be seen from late August to early October, when both adults and young bats, born a few months earlier, are flying. During the summer season, a Ranger gives a short talk before the flight begins. As cold weather approaches, the bats desert the cavern and migrate south to Texas and Mexico for the winter. Some banded bats from Carlsbad Cavern have been recovered in Mexico more than 800 miles away.

In recent years, for reasons that are not clear, the bat population has been declining. Most of the bat's life is spent outside the boundaries of the park. They feed all night outside the cavern in summer, and fly south in winter, so the Park Service has little control over many of the important circumstances in the bats' lives. The bats feed on flying insects, and anything that affects their food—either in the immediate area of the cavern or in their winter range to the south—obviously may affect the bats. Probably many of the former habitats which provided a large supply of insects have decreased. More and more of the desert is being utilized for farmland and housing developments. Even more important is the increased use of pesticides, which has affected the bats both by decreasing their food supply and by acting directly on the bats themselves. Poorly understood epidemic diseases also may take a toll. In Mexico, where vampire bats with rabies have been found, some bat caves have had almost their entire populations destroyed, even though the vampires were just a small fraction of the bat population. Other means of control are now being sought.

Over the years Rangers noted that the water level in most of the pools in the cavern was falling. In 1968, a joint U.S. Geological Survey-National Park Service project was initiated to try to determine the reason. It was found that the principal cause of the

The bats circle about for several minutes after flying out of the cave. *National Park Service*.

drying was excessive airflow up the man-made elevator shafts, which acted as 800-foot chimneys during the dry winter months. A secondary factor was the heat generated by the cavern lighting system. Surface developments and the large numbers of visitors appeared to have a less significant effect. The net loss of water through the elevator shafts was estimated to be about 22,000 gallons per year. To prevent further losses, revolving doors were placed in the lower elevator lobby to provide an effective seal, and the cavern is being rewired with more efficient and less heat-producing fluorescent and mercury vapor lights. A careful program to refill depleted pools was undertaken.

Equally careful attention has been given to the surface facilities. The visitor center provides a restaurant, gift shop, and small museum. Even baby-sitting and kennel services are available. In an attempt to maintain the area in as nearly a natural condition as possible, surface facilities such as administrative offices and ranger housing have been, or will be, moved outside the park boundaries. A new sewage disposal system with retention lagoons below the escarpment has been built at a lower elevation, moving such facilities several miles from the cavern. The present road up Walnut Canyon from the park entrance to the cavern is narrow. It is the only road to and from the cavern and bears a heavy load of two-way traffic. The scenic pleasures of the drives are often negated by the heavy traffic of all types. A new exit loop which would allow only one-way traffic up Walnut Canyon has been proposed.

An experience second only to true spelunking is available in another of the park's caves. Primitive lantern tours of New Cave, the second largest cave within the park, are led by Rangers and provide visitors a different type of caving experience. New Cave contains a wealth of speleothems, some quite large and beautiful. But it seems unlikely that this or any other new attraction in the park can begin to rival the old giant.

As more and more people visit the cavern it may become necessary someday to limit the number of visitors permitted inside at the same time. Caves are fragile and can easily be damaged irreparably, even by well-meaning individuals. A single footprint in the wrong place may leave permanent dam-

age, and a moment's carelessness can destroy a stalactite forever. Only through careful attention and constant surveillance can Carlsbad Cavern continue to be a unique national resource.

A caver views the Christmas Tree in New Cave. *National Park Service*.

Survival: A Tale of Two Bats

The true facts about bats are fascinating enough; it is unfortunate that there are so many misconceptions and untruths of folklore about them. Bats, being mammals, are warm-blooded, and their young are born alive. They are the only true flying mammals. (Flying squirrels glide, but do not really fly.) Most bats are active at night, sleeping during the day in caves, buildings, or trees. They have functional eyes but navigate through sonar or echo-location: the bat emits high-pitched squeaks or cries which echo off nearby objects and orient the creature. All bats in the United States are insect eaters; they catch the insects on the wing through their sonar system. Vampire bats do exist in tropical America, and can be health hazards—not so much for the blood they drink, but for the diseases they may carry and transmit through biting. Other kinds of bats eat fruit, or even catch fish.

Bats often use caves for hibernation or maternity roosts as well as for daily sleeping quarters. Some southwestern caves harbor millions of bats. Eastern caves usually have smaller populations. Naturalist-speleologist Charles E. Mohr has been studying these creatures for more than forty years. During that time the bat populations of most caves have shown significant decreases, and several species, almost wiped out, have been placed on the List of Endangered Species.

Mohr's interest in and contributions to speleology extend back to the 1930s. In addition to his biological expertise he is a leading photographer of caves and cave life. He has written several books and scores of articles about the many facets of his work, and his pictures have been widely published. He is both a former president of the National Speleological Society and an honorary member.

Survival: A Tale of Two Bats

CHARLES E. MOHR

When I plucked two kinds of bats from the ceiling of Tennessee's Nickajack Cave west of Chattanooga, I could not have guessed that forty-five years later both would be on the United States Department of the Interior's List of Endangered Species.

It was on the day before Christmas in 1931 that I checked out the belief of U.S. National Museum mammalogist Arthur H. Howell that gray bats occupied Nickajack and other caves the year round. He had collected many specimens here in 1908, expecting them to be *Myotis velifer*, a species quite common farther west, and the largest of a dozen kinds of *Myotis* (bat) then known.

What Howell found was a new bat, the size of *velifer*, but grayer and with the wing membrane reaching only to the ankle, not to the toes as in other related bats. He named it with the latin word for gray: *grisescens*.

Guided by a map published in 1881, Howell had entered the

Speleologists Dorothy Reville and Howard N. Sloane observe Indiana bats, *Myotis sodalis*, in Longs Cave, Mammoth Cave National Park, Kentucky, ten years before this species was placed on the U.S. List of Endangered Species. Protected by a cave gate and restricted access, this colony is surviving. *Charles E. Mohr.*

cave alone in August 1908. He found his way by torchlight for nearly a thousand feet alongside one of this country's largest underground rivers. He then fired a single shot from his double-barrelled shotgun into a mass of excited bats hanging overhead—a collection method no speleologist would use today. Evidently the vibration stunned more than were hit. Scores fell, many into the river. Most of those quickly swam away, but he was able to collect seventy of the fallen bats.

In 1931, my brother John and I had a guide, Dr. L. S. Ashley, operator of the cave. We soon found the blackened section of ceiling where Howell had shot into the cluster of gray bats. Now, of course, they should be dormant. But we found no bats here. Ashley was not surprised. He said that in winter the bats always hung on the far side of the stream. It was sixty feet wide—deep, cold, and uninviting.

Converging walls almost forced us into the water, but 200 yards ahead the cave widened, and here we found several hundred bats, within reach on the low ceiling. These weren't gray bats. They were brown, with a purplish cast. I recognized them to be a smaller species, *Myotis sodalis*, which I had seen for the first time just a few months earlier in Pennsylvania.

This was a species of bat virtually unknown even to bat experts because it had not been recognized as different until 1928. Only one specimen had been found in the state of Tennessee. Here were hundreds. There were thousands across the stream, Ashley said, so when he pointed out a shallow place where he said the water was only four feet deep, I stripped and waded in. It was just chin deep—but a bone-chilling 58 degrees Fahrenheit.

Upstream the rock wall rose from the water's edge. Downstream huge blocks had fallen from the ceiling, creating a maze. Here I found four clusters of bats—about 1,500 *sodalis*—like dark rugs on the ceiling. Three solitary bats proved to be *grisescens*. Where were the thousands I had seen in summer?

Decades passed before we had the answer, when the travels of the gray bat could finally be plotted on a map, but in 1931 we determined that Nickajack Cave sheltered two bat populations, not one. In winter it was *Myotis sodalis*, and in summer only *M.*

grisescens, as we verified the following June.

The *sodalis* were gone by then. I hadn't expected to find any, because in Pennsylvania I had inspected their wintering caves and found the bats gone by May. Where did they go? For four decades that was possibly the most puzzling question for students of bats—the disappearing act of the Indiana bat, *Myotis sodalis*. Its summering grounds were still unknown when, in 1966, it was placed on the U.S. Department of the Interior's Endangered Species List, along with the Hawaiian hoary bat, *Lasiurus cinereus semotus*, a tree-dwelling species.

How are "new" bats discovered—recognized as different from any species previously known? Seldom is the collector himself expert enough to correctly identify every bat. Even when specimens reach a museum they may be incorrectly identified. That happened with the first specimens of both *M. grisescens* and

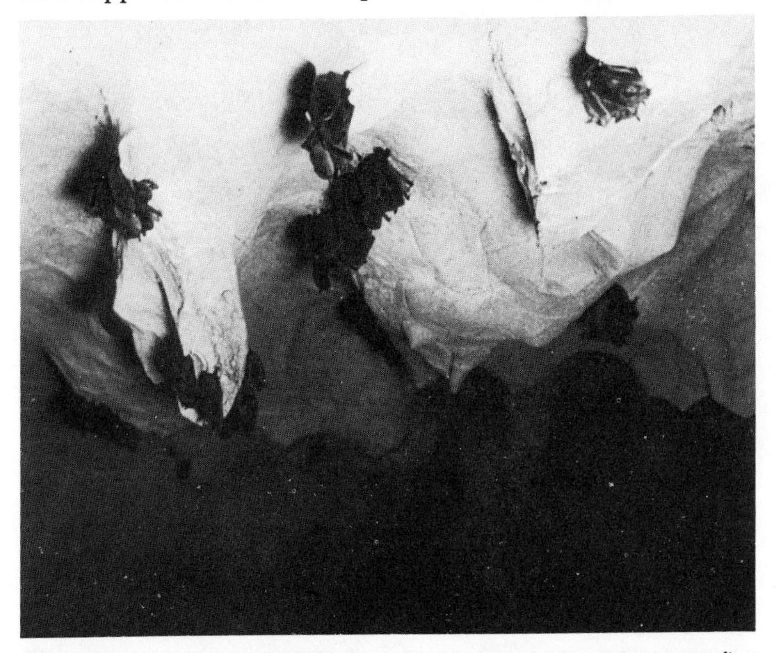

Photographed in 1932 in Nickajack Cave, Tennessee, where they were first discovered, the gray bat, *Myotis grisescens*, was added to the list of Endangered Species in 1976. *Charles E. Mohr.*

M. sodalis. It may not be until years later that the specimens are studied by a bat specialist and their uniqueness determined.

As a biologist interested in cave animals, I had gone to Nickajack looking for a green-blotched salamander, *Aneides aeneus*. It was discovered there in 1881 by the noted paleontologist Edward Drinker Cope and his zoologist friend A. S. Packard. I was also hoping to find the handsome blind white crayfish which they had found a mile inside the cave, as well as an undescribed white isopod, and an eyeless white amphipod named as new by William P. Hay in 1903.

No one who has been to Nickajack will ever forget it, even today after the Tennessee Valley Authority's newest impoundment, Nickajack Lake, has submerged half of the cave's sixty-foot-high entrance. Its mammoth flat span measures 160 feet. Illuminated by the early morning summer sun, this is one of the

Too easily accessible in 1932, Nickajack Cave is now entered only by boat; a TVA dam forms Nickajack Lake which raises the water level halfway to the cave's roof. Both protected species of American cave bats use the cave, one for hibernation, the other for a maternity roost. *Charles E. Mohr*.

most impressive cave entrances in the land. And it is visible from the highway, a half-mile away.

Hay, too, had been looking for invertebrate cave fauna in 1903, but he collected two bats. These were the specimens which Howell identified as the western *Myotis velifer*, possibly with some doubt. At any rate, he decided five years later to investigate the Nickajack bats for himself—and wound up describing a new bat.

My own observations that Christmas Eve in 1931 inspired a return visit in June. We took along a colleague from the Reading, Pennsylvania, Museum, Kenneth Dearolf, who, like William Hay, was a specialist on cave invertebrates.

Since most of the cave animals—small, blind, and white—are aquatic, we concentrated our efforts on caves known to have streams running through them. As it turns out, both *grisescens* and *sodalis* are water-oriented, the former spending summers along lakeshores, the latter along stream valleys, a fact unknown until 1973.

On the banks of the Holston River in east Tennessee we found a nursery colony of *grisescens* in beautiful Indian Cave. While Dearolf followed the cave guide along the underground stream, John and I climbed to an attic-like chamber close to the entrance.

It wasn't simply the exertion that caused us to sweat as we explored the bat chamber. The air temperature here was ten degrees higher than in the cool cave passage below—the result of thousands of warm bodies and the chemical reactions within the mounds of fresh bat guano, wet with urine. Only in the caves of the freetailed bats of the Southwest have I experienced stronger ammonia fumes.

Several thousand gray bats hung from the walls and ceiling, young bats scattered among the adults. Within minutes hundreds of bats were flying. Bats hit us repeatedly. There were far too many bats flying for the amount of air space. Baby bats, dislodged from their ceiling roosting sites, were crawling across the guano-covered floor. Probably they were doomed by the host of predacious beetles and mites that populated the fresh guano.

After a hasty retreat from the bat room, we learned from the manager that for more than two weeks he had been removing young bats "by the shovelful" from the floor of the cave below the bat room. Mothers leave the baby bats hanging from the ceiling as they go out at night to feed. Returning, the mother bats find their own babies, evidently recognizing them by individual vocal signatures—part of the audible distress calls given continuously when the young are separated from their mothers.

The youngest juvenile we found was completely naked and weighed just 2.5 grams. The largest had a heavy covering of fur, weighed 5.4 grams—about one-fifth of an ounce—and probably was within a week of flying. Most gray bats fly by the time they are four weeks old.

The next day we were back at Nickajack, this time with a boat. We found the winter-empty roost sites repopulated by thousands of active gray bats. This time we explored the underground river to its upstream source, a deep pool, without finding any more gray bats. And now in June, there were no *sodalis* whatever. The winter colony had vanished.

A few days later, June 28, in Mammoth Cave National Park, we revisited Dixon Cave. Where 2,500 *sodalis* had wintered we found none until we climbed to the top of a breakdown rock pile at the end of the 800-foot-long trunk passage—known to be an extension of Mammoth Cave. Here we found one compact cluster of torpid bats. We examined them one by one. All were *sodalis* and all were males. We had no idea where the females were. Until John S. Hall began a four-year doctoral study of *M. sodalis* in 1956, this was the only solid data on the summer whereabouts of the species.

From the common name, Indiana bat, for *Myotis sodalis*, one might imagine that Indiana is where the species is most common. It is not, but Kentucky is. It just happened that when two bat experts, Gerritt S. Miller, Jr., at the U.S. National Museum in Washington and Glover M. Allen at Harvard, were methodically examining collections during the mid-1920s, some rather strange bats were found in jars of preserved little brown bats, *Myotis lucifugus*. The little brown bat is the commonest *Myotis* in the northeastern states. Though similar in size to *lucifugus*, the other bat's fur, when dry, was different: light gray (never glossy or yellowish) at the hair's tip, dark gray along most of its length, and almost black at the base—tricolored hair.

Viewed in a cave by carbide lamp or flashlight, the fur seems to have a purplish look. And in the great masses that cling to the walls and ceilings of Bat Cave in northeastern Kentucky, the pale lips of *sodalis* make them easy to distinguish from the dark-faced little browns that hang loosely in the colder entrance chamber.

A solitary bat, however, is difficult to identify, particularly a dead one preserved in alcohol or formalin. When Miller and Allen separated the specimens with tricolored fur from the familiar *M. lucifugus* and examined the skulls, they found an additional feature—narrow, more delicate skulls, usually with a slight crest down the center.

Some of these bats had been collected twenty years earlier, but of course had not been recognized. Miller and Allen chose the bat with the earliest collection date to be designated the "type specimen." It just happened to come from Wyandotte Cave, Indiana. Others, nearly as old, came from Pennsylvania, Vermont, and Kentucky. The choice of the name "Indiana bat" was unfortunate, in my estimation. The scientific or latin name selected was much better—*sodalis*, which means "companion." Field investigation of the new species had revealed its clustering behavior to be its most distinguishing characteristic. When the National Museum published a complete bulletin on bats of the genus *Myotis* in 1928, the single new species reported was *Myotis sodalis* Miller and Allen.

Two years later, while I was a graduate student at Bucknell University in central Pennsylvania, I consulted this bulletin and read that specimens of *Myotis sodalis* had been collected in "a cave in Central Pennsylvania." I knew that there was a commercial cave at Woodward, just thirty miles to the west, so I hurried there in hope of finding the new bats. They should be hibernating there at this particular time.

With the cave's lights turned on it was easy to spot the dormant bats. Just inside the icicle-draped entrance were some good-sized, pale chocolate-colored bats which I recognized as big brown bats, *Eptesicus*. Farther in, where the cave thermometers registered a steady 56 degrees Fahrenheit, I found scattered tiny yellowish pigmy bats, *Pipistrellus*. And clustered among the stalactites and scattered on open wall surfaces were medium-sized *Myotis*, which I assumed were the new species, *Myotis sodalis*.

With the owner's consent I picked off a dozen *Myotis*, dropped them into a bag, and carried them back to the campus in my car's rumble seat. Later, when I lifted the lid, wide-awake bats flew out and away. The trip wasn't a total loss, however. There were three bats still in the bag, so I took them to the Reading Museum and gave them to curator Earl L. Poole, a man who had had a lot to do with stimulating my interest in natural history.

It took him only moments to determine that my specimens were not *M. sodalis*. Two were ordinary little browns, *M. lucifugus*. The third was smaller, its fur more golden, its ears, lips, and wings quite black—a bat he had never seen before; unquestionably it was *M. leibii*, named in 1841 by John James Audubon and the Reverend John Bachman. Only eight specimens existed, as we learned when we read the Miller and Allen bulletin.

Two additional species during hibernation, big brown bats, *Eptesicus fuscus* (left) and the smallest species, *Myotis leibii*, share a Pennsylvania cave. *George F. Jackson.*

"How many do you want?" I asked curator Poole. "The cave is full of them." It wasn't—no more, at least that winter. But in further searches for them I visited the right cave, unnamed in the bulletin, and found 2,000 *sodalis*. It turned out to be the popular show cave, Penns Cave, near State College; it was seldom visited in winter, when the clusters of bats hung forty feet above the deep river, on which large boats carried sightseers on a quarter-mile round trip in season.

I soon found a second colony of *sodalis*, about 500 bats, in yet another commercial cave near Bedford, Pennsylvania; this time they hung twenty feet above a shallow cave stream which separated the bats from the tourist path. Even minimal winter visitation here constituted a threat to the bats.

A third colony of *sodalis*, also about 500, shared the chambers of undeveloped Aitkin Cave, near Lewistown in central Pennsylvania, with more than 4,000 *lucifugus*. Unfortunately for the bats, Aitkin was destined to become the most popular spelunking cave in the area.

Within twenty years disturbances by commercial cave or spelunker traffic drove the three *sodalis* populations from Pennsylvania, although it may have been a near-record flood in 1950 that wiped out the Aitkin cave population. Bat expert Donald L. Griffin and his party were nearly trapped by that Thanksgiving weekend deluge. At Big Spring in Kentucky a flash flood in March 1964 drowned an estimated 4,000 *sodalis* and 2,500 *lucifugus*.

John Hall believes that *sodalis* is especially vulnerable to floods. In spring and fall great numbers of bats appear briefly at many caves along the Green River in Kentucky, and along other river valleys, evidently making regular stops resembling the visits of migratory waterfowl to wildlife refuges along their migration routes. A number of these river valley caves also are used for hibernation. In Bat Cave, seemingly safely high above the Green River in Mammoth Cave National Park, Hall discovered that earlier explorers digging through a crawlway had exposed a bone deposit twenty feet long, two to three feet wide and three to four inches deep. In just eighteen cubic inches Hall counted bones and skulls from 200 bats, and calculated that 300,000 bats—*sodalis*—had died here in one catastrophic flood.

This colony was three times larger than any *sodalis* colony known today. Its fate underscores the vulnerability of the species. For reasons not understood, almost the total wintering population of *sodalis* is concentrated in four caves. Was it five, before the flooding of Bat Cave?

Since 1938, when W. A. Welter of Moorehead State College introduced me to the enormous colony wintering in another Bat Cave (in Carter Caves State Park in northeastern Kentucky), I have made a visit to that cave part of any winter trip through the midwest. Even though it was accessible to the public, and though floods flushed through the hibernation chambers every few years, the population, which I estimated at 125,000 in 1938, seemed to be holding its own remarkably well—until 1957.

During three Christmas school vacations between 1957 and 1960, high-school students entered the cave, stoning the bats, bombing them, and knocking thousands of them into the cave stream to drown. A huge gate, hinged to permit it to swing up out of the way of flood waters, was built by Ralph Ewers of the Cincinnati Museum of Natural History with the help of concerned spelunkers, but it failed to keep vandals out.

Finally in 1969 the Kentucky State Park Commission erected chain-link fences around the two entrances. On an inspection trip in March 1975, John Hall and Stephen R. Humphrey estimated that the population had declined to 30,000 bats.

The only colonies of *sodalis* that have been successfully maintained are the ones in Mammoth Cave National Park. Colonies which I first saw in 1931 in Dixon and Long's Caves seem to be identical in size today. What to me is most remarkable is that the clusters in 1975 were precisely at the spots where I found them in 1931 and half a dozen other times in the intervening forty years.

In one room in Long's Cave their distribution is unique; they hang from extruding edges of calcite crossing the flat limestone ceiling in a random pattern. Everywhere else *sodalis* hangs in the characteristic shoulder-to-shoulder, upside-down carpet on the ceiling.

The difference protection makes impresses me as I look over

counts I made in years past, compared with recent estimates, mainly by Hall, from 1972 to 1975, as summarized in the accompanying table.

BAT POPULATION CHANGES IN EASTERN U.S. CAVES

CAVE	Original Survey		Latest Survey	
	Year	Number of Bats	Year	Number of Bats
	DECREASED BAT POPULATIONS			
Wyandotte, Indiana	1932	15,000	1974	1,900
Bat, CCSP,* Kentucky	1938	125,000	1975	30,000
Coach, Kentucky	1959	100,000	1975	4,500
Colossal, MCNP,† Kentucky	1931	2,000	1975	14
Penns, Pennsylvania	1931	2,000	1952	0
Hipple, Pennsylvania	1931	500	1952	0
Aitkin, Pennsylvania	1931	500	1952	0
White Oak Blowhole, Tennessee	1950	8,000	1974	6,050
Starr Chapel, Virginia	1960	1,800	1974	30
Trout, West Virginia	1935	4,000	1972	0
	INCREASED BAT POPULATIONS			
Long's, MCNP,† Kentucky	1931	2,000	1975	7,600
Dixon, MCNP,† Kentucky	1931	2,500	1975	3,600

*Carter Caves State Park
†Mammoth Cave National Park

The single hopeful sign I read from these statistics is that total protection, as in Mammoth Cave National Park, appears to hold some promise for relocation of harassed bats. Long's and Dixon Caves are good examples. The decline at Colossal Cave resulted from a complete barrier gate that excluded the bats. It appears that the bats shifted to Long's Cave, a mile away. That raises a question: could a population be physically transferred from a cave scheduled to be inundated, destroyed, or made untenable by commercial development?

A number of important populations of cave bats in Arkansas, Missouri, and Alabama face dispossession. Construction, long planned in the U.S. Forest Service's pre-eminent cave wonderland at Blanchard Springs in Arkansas, has dispossessed several thousand *sodalis*. Since other populations spend the winter in nearby caves, it is hoped that the expatriated bats will join them, that enough additional completely suitable areas exist there to accommodate additional populations.

In the Meramec Valley in Missouri, construction of the Meramec Park Lake will probably flood 140 caves, including twelve known to be important bat caves. Under what circumstances might the dispossessed bats find suitable alternate quarters? We need to find out.

To comply with federal requirements for an adequate Environmental Impact Statement, the U.S. Army Corps of Engineers is funding an eighteen-month study of cave bats by a team of biologists from the University of Missouri, in cooperation with the U.S. Fish and Wildlife Service and the Missouri Department of Conservation.

The research team, headed by Richard K. LaVal, has already pinpointed the dozen largest populations, including *Myotis sodalis* and *grisescens*, and four other species of *Myotis*. In its survey, the team traps and bands up to ten percent of the bats in flight as they arrive at the hibernation sites. During the winter, periodic, quick monitoring will determine the stability of cave populations and any intercave movements. Combinations of colored bands on either wing will permit instant detection of such shifts and may reveal the likelihood of self-initiated, successful relocation by expatriated species.

At the moment there seems to be little basis for the optimism expressed in the "Recovery Plan for the Indiana Bat" recently completed by a U.S. Fish and Wildlife Service team of wildlife experts and bat biologists. The report predicts a comeback—including repopulation of the abandoned hibernacula in the Northeast, from Pennsylvania to Vermont.

My own effort to move a little brown bat nursery colony from a dwelling house in Pennsylvania to an abandoned house nearby was a total failure. Pregnant females displayed an astonishing loyalty to the original roost and total disdain for the proffered alternative.

Even more relevant to the Meramec situation is the case of several thousand *Myotis lucifugus* which we carried away from one of the old railroad tunnels as it was being enlarged and holed through for the Pennsylvania Turnpike in 1940. We released the bats in the state's biggest bat cave and they evidently spent the rest of the winter there.

But the next fall they reappeared at the site of the old tunnel, flying back and forth in the newly completed highway tunnel, vainly seeking their dark, quiet retreat with its narrow range of humidity and temperature—the microclimate which permitted them to metabolize their fat reserves slowly enough to last out the winter. The transfer didn't work.

The Meramec bats won't lack alternate cave sites. There are

On-the-spot studies of internal temperatures (ear and rectum), of the microclimates of different clusters and shifts of hibernating locations—carried on with a minimum of disturbance—provide data which may be critical to the survival of the gray bat, *Myotis grisescens*. (Study terminated when species was placed on Endangered List.) *Merlin D. Tuttle.*

at least one hundred caves which would be above dam level in the area under study. The catch is that each species of *Myotis* has a remarkably narrow set of microclimate requirements—parameters of temperature and humidity—that must be met to insure an economical metabolic rate. A bat occupying a microclimate unsuitable for its species will arouse more frequently than is safe during winter hibernation.

In some caves, areas of seemingly optimum temperature may be exposed at times to air currents that critically alter the microclimate. A species like *M. grisescens,* which selects the coldest, usually the deepest, caves for hibernation, can be adversely affected by any structural change that reduces cold-air circulation in winter. Just outside Mammoth Cave National Park, one of the three or four most important wintering populations of gray bats occupies the lowest portion of a cave recently commercialized.

While the owners cooperated by suspending trips to the bat area in winter, they built a gift shop over a natural airshaft to take advantage of the natural air conditioning—as half a dozen other commercial cave operators have done. But this Kentucky operation affects bats. The temperature deep in the cave is now higher than the gray bats or a colony of Indiana bats can tolerate for long. Even if the gray bats survive the winters, they may be too weak to make the exhausting flight to the caves where they spend the summer, possibly several hundred miles away, according to Merlin D. Tuttle of the Milwaukee Public Museum.

In 1974 Tuttle completed a fourteen-year study of the gray bat, and, largely on the basis of his findings, *Myotis grisescens* was added to the List of Endangered Species by the U.S. Fish and Wildlife Service in 1976.

While my observations at Nickajack Cave in 1931 showed that *grisescens* used that cave only in summer, and the 14,500 bats that I photographed in Marvel Cave, Missouri, in 1935 were there only for hibernation, the obvious shifting of cave scenery was perplexing, to say the least.

It took the most intensive study of cave bats ever made to reveal the amazing saga of the gray bat, a species that sometimes congregates in colonies numbering a quarter of a million, a species that occupies certain deep caves that lure those spelunking daredevils the vertical cavers, a bat that in summer prefers caves along river valleys where water-born insects are abundant but where engineers envision mammoth reservoirs. To top it off, several of the big caves occupied by *grisescens* colonies are spectacular enough to warrant development as tourist attractions. The gray bat is really in trouble.

The seasonal shifts are now known to involve flights long enough to be hazardous. The physical stress may be considerably increased during periods of stormy or unusually cold weather, when insect activity may be markedly reduced. Gray bats migrating northward from one Florida cave in spring usually deviate from a straight-line distance of 260 miles over noncave country in favor of a safer 350-mile route mostly along river valleys, that keeps them closer to limestone caves.

Tuttle's study of thousands of repeat recaptures, including more than 500 records of round trips between summer and winter caves over a fourteen-year period, made it possible for him to plot migratory routes with certainty and to demonstrate remarkable "loyalty"—philopatry—on the part of individual bats to their long-inhabited nursery or hibernation caves.

Study of the nursery caves used by *grisescens* revealed two essential criteria: the caves needed to possess physical size and configurations that favored the retention of maximum temperatures, and they needed to be close to bodies of water with extensive shallows where populations of flying insects were large. Tuttle's statistics for successful growth and development of young bats show dramatic declines as distance of the maternity cave from water increases.

Evening feeding flights from caves near or opening onto T.V.A. impoundments such as Nickajack Lake are spectacular and may indicate conditions favorable to an increase in gray bat populations.

Access to most of these caves is difficult, and with wider public recognition that the new regulations will be strictly enforced in federal courts, it is hoped that acts of vandalism against bats are likely to be much reduced or nearly eliminated.

The situation with *sodalis,* the social or Indiana bat, is quite different. In 1973 and 1974 investigators in Missouri, headed by Stephen Humphrey of the Florida State Museum and James Cope of Earlham College, finally found maternity roosts. Instead of moving *en masse* into a few nursery caves as *grisescens* does, *sodalis* scatters along river valleys and forms numerous but obscure colonies sheltered under loose bark and in tree cavities. The river-edge trees are mainly sycamores and cottonwoods.

Foraging habitat is restricted to open space near the foliage of these river-bank or riparian trees, and to a few isolated trees in pastures on the creek floodplain. When the weather warms and the young are born, feeding habitat is expanded to include forest-edge trees on the floodplain.

Natural destruction of such tree shelters is normally on a scale small enough to permit successful shifts to alternate roosts, already familiar to the bats. Of much greater consequence, however, is the wholesale destruction of riparian habitat resulting from dam construction or stream channelization, both of which are popular with the U.S. Army Corps of Engineers. At present no movement of banded *sodalis* from caves to specific maternity tree roosts has been reported, but the likelihood of discoveries resulting from the present Meramec Lake Park study is indeed promising.

Virtually everyone involved in bat research and protection agrees that second only to rigorous enforcement of federal Endangered Species regulations is the need for acquisition of caves where major wintering and maternity populations are concentrated. Only a few are now owned by federal or state agencies. The Nature Conservancy owns several caves identified as habitats for endangered species of wildlife. They, and The Ozark Underground Laboratory, a cave owned by speleologist Tom Aley, are notable but exceptional landmarks. Federal and state recovery plans propose acquiring critical caves, as authorized by the Endangered Species Act of 1973, but at this writing no funding has been authorized.

Exploration of the Tanama

Rio Tanama shows up clearly on aerial photographs or from an airplane. The river twists and turns through the jungle of northern Puerto Rico, often hidden by the lush vegetation. But in some places it disappears completely—only jungle and rocks are visible—and flows underground for several thousand feet or more.

Russell Gurnee knew there was a good possibility of one or more big caves. He had already explored the nearby Rio Camuy, finding the largest and most extensive cave in Puerto Rico. Did the Rio Tanama hold equal promise? With fellow members of the Explorers Club he went to find out.

Gurnee has served as president of both the Explorers Club and the National Speleological Society, and has led caving expeditions throughout the world. His earlier explorations in Puerto Rico are recounted in *Discovery at the Rio Camuy*, written with his wife Jeanne. The following article first appeared, in a slightly different version, in *Explorers Journal*.

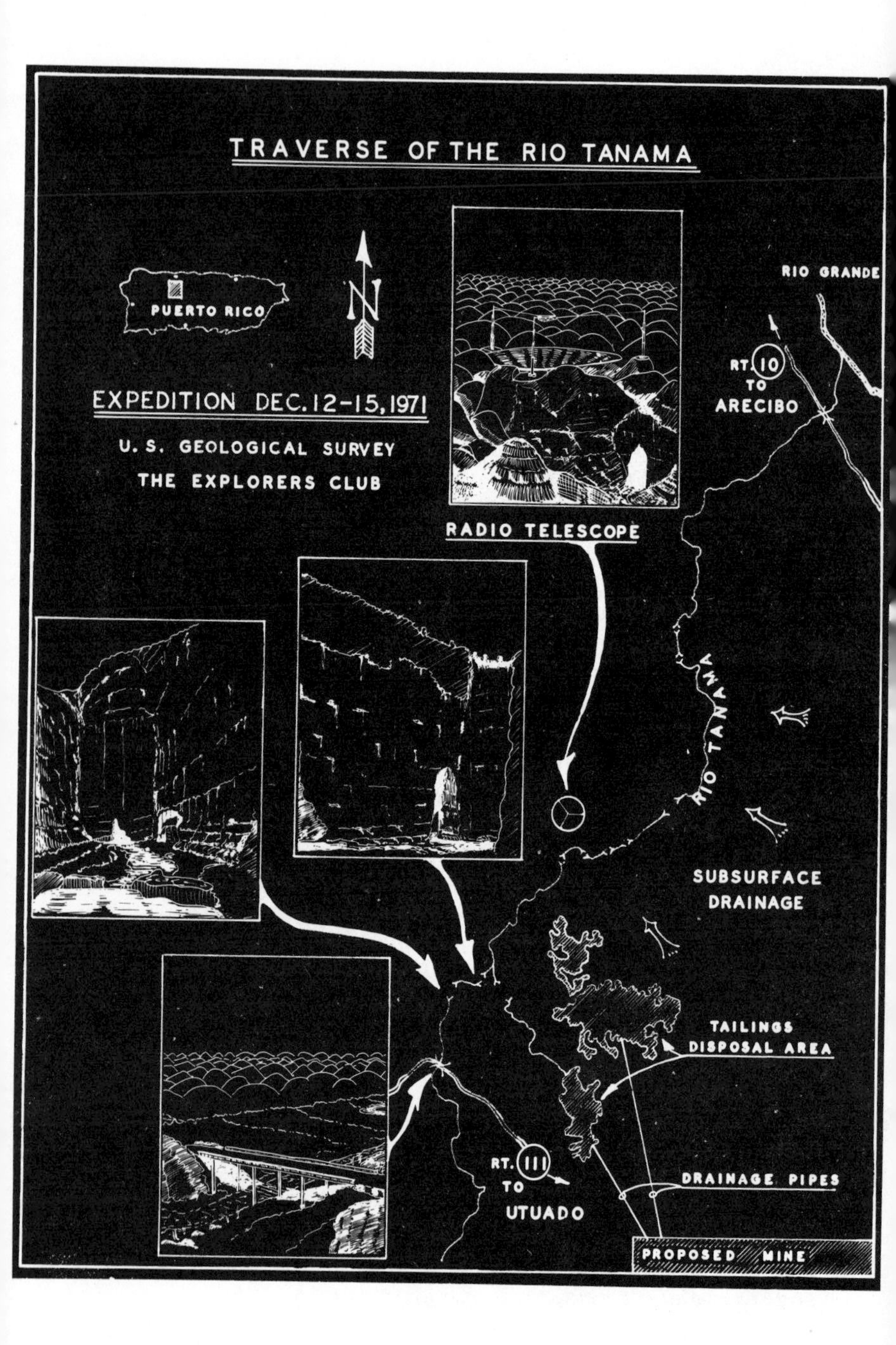
TRAVERSE OF THE RIO TANAMA
PUERTO RICO
N
EXPEDITION DEC. 12-15, 1971
U. S. GEOLOGICAL SURVEY
THE EXPLORERS CLUB
RADIO TELESCOPE
RIO GRANDE
RT. 10
TO
ARECIBO
RIO TANAMA
SUBSURFACE
DRAINAGE
TAILINGS
DISPOSAL AREA
RT. 111
TO
UTUADO
DRAINAGE PIPES
PROPOSED MINE

Exploration of the Tanama

RUSSELL H. GURNEE

Rivers have always been the highway for the explorer, as they provide a natural trail into a previously unexplored and unmapped area. Occasionally natural barriers in the rivers have prevented their exploration; but with the advent of aerial photography and mapping, it is rare to find an "unexplored" river left to traverse.

One such river was the Rio Tanama—a fair-sized watercourse in the sinkhole karst area of Puerto Rico. It would seem that an unexplored river would be unlikely in a location only fifty miles west of San Juan and a few thousand feet from the National Astronomy and Ionosphere Center, where the largest radio telescope in the world is located.

Attention had been focused on the river in 1966 when copper-bearing ore was discovered near Utuado, several miles to the southeast. The United States Geological Survey made a study of the river and did a series of dye tests which showed a direct connection from nearby sinkholes to the river water, but it was not possible to determine whether the water entered through springs or caves that might provide physical access.

If a more accurate determination of the underground flow of

Location map, Rio Tanama, Puerto Rico. *Russell H. Gurnee.*

water were to be made it would be necessary to travel the length of the river (approximately thirteen miles), to investigate the caves along the route, and to sample the springs that were found.

There were many problems involved in making such a trip, because the area is virtually uninhabited, with rugged topography. Although the relief is only three or four hundred feet from the tops to the bottoms of the "haystacks" (cone-shaped limestone hills), there are only a few foot and horse trails. The only roads are to the telescope and to a point near a purification plant en route. The only surface stream is the Tanama, which has become the drainage trunk for all the water that falls and flows into it from the Central Cordillera to the south. The average rainfall exceeds eighty inches per year, and flash floods are common.

Study of the air photos of the river showed at least five places where the river flowed underground. In many places the river was covered with trees, making it impossible to determine whether there were other underground channels. One thing was sure: the river dropped over eight hundred feet in its thirteen-mile trip, and there were certainly several places where it plunged in waterfalls, visible from the air. It was not possible to tell whether we would be able to portage around the falls and rough rapids. It would be necessary to go and explore.

In December of 1971, three members of the Explorers Club—Roy Davis, Addison Austin and I, plus Jack Herschend, and Don Jordan and Vito Latkovitch from the United States Geological Survey—made this interesting traverse.

Our trip began at the bridge, Route 111, at Angeles, Puerto Rico. There were ten of us on the gravel bar—six of the crew to go down river and four of the support party headed by our friend, Watson Monroe, who would meet us at the radio telescope to pick up water samples and to supply us with additional food. We piled all the gear in one huge mound and glumly surveyed the river. It was only about twenty feet wide; as far as we could see downstream, there appeared to be rocks breaking the surface.

"There will be more water downstream," Vito mentioned. He pointed to a water gauge on a cement pylon of the bridge. "It is flowing about thirty gallons per second now. Last month the high water mark on the gauge was twelve feet above the present level, and it came up in less than two hours."

This was hardly reassuring, for we were hoping to avoid the rainy-season flooding. But we needed more water than we could see in the river if we were to make a successful passage.

Vito spoke with authority, as he had been studying the flow of this and other rivers in Puerto Rico for the past year to measure the drainage of the region. As part of his work with the Geological Survey in Puerto Rico, he chained a recording box to the bridge; it gave hourly readings of the flow of the river.

"The river rises and falls very quickly," added Vito. "Right now there are probably one hundred gallons per second at the other end. We know of three springs and tributaries that add to the flow. How many more there are we will have to find out."

The explorers make a swift deep passage through large blocks of limestone. *Roy Davis*.

We dumped out the rubber boats from their seabags, and, using a foot bellows, began to inflate them. There were two heavy-duty black landing craft and a lighter yellow boat.

The boats were inflated and tossed into the water with bow and stern lines secured so that they could be controlled from either end. Jack Herschend and Add Austin were the most experienced river men, so I suggested they each take the stern of one of the heavier boats; Don and Vito from the USGS were in the yellow one. Roy Davis cheerfully went into Add's boat muttering something about my being Jack's handicap.

As we launched, we agreed to meet the support party in two days at the base of the trail from the radio telescope, to receive more food supplies and have the water samples taken out.

At the start there was not enough water to float the heavily laden boats, so we got out and pushed and tugged the boats ahead and around the bend, out of sight of the party at the bridge. Fortunately soon there was a pool and a stretch of calm water, and we had an opportunity to paddle another hundred yards before we had to disembark and walk.

We found that we could walk along the shore and, with a long tether, make good time as the boats drifted over the gravel. This first part of the river flowed through comparatively flat land. Bamboo and sugar cane grew down to the water, with small trees providing a screen from the highway. The stream was strewn with small volcanic boulders tumbled and rounded by the swiftly flowing water. There was no brush or debris blockage and no trees or branches to snag the boats—just not enough water.

We were not making very good progress. It was into the boats and out again, wet to the waist until a sudden shower gave us a complete baptism. Since it was tropically warm and mild, our spirits remained high.

Several small springs by the edges of the stream caught the attention of Don and Vito, and we stopped to take temperatures and fill sample bottles. The sky cleared; the sun appeared, and it suddenly became warm. I looked at my watch: it was nearly 1:00 P.M., so we stopped for lunch.

Later we were again on the river and saw hills about a

quarter of a mile ahead. The banks of the river were now too steep to climb, and the stream ahead was littered with limestone blocks that had fallen into the canyon. With the constriction of the walls, the water became deeper, and progress was better. In another hour, the canyon hooked sharply to the right and stopped at a sheer 400-foot wall. The river plunged ahead, cascading down through the rocks and disappearing into a broad cave opening at the base of the cliff.

We gathered on the gravel bar in front of the cave and surveyed our situation. The canyon was about a hundred feet wide at this point; directly ahead was a twenty-foot-wide opening that was undoubtedly the way to go.

Off to the right of this entrance and ten feet above the water was another cave—this one dry and possibly an ancient course for the river. Since our mission was to explore the possible watercourses that supplied the Tanama and since this was the first cave we had seen, we got out flashlights and carbide lamps and prepared to make a reconnaissance.

Roy Davis was the first to be ready, which was to be expected as he is undoubtedly the best and most experienced cave

Preparing to enter the first underground section of the Tanama River. The end of the tunnel is visible as a pinpoint of light inside the cave opening. *Roy Davis*.

explorer I know. He has spent more than twenty years as a consultant, lighting expert, and cavern developer throughout the United States.

Jack and I waded across the river after Roy; while they disappeared into the lower crawlway of the downstream passage, I started to climb up a steep mud slope toward a high upper passage that seemed to parallel the canyon, going upstream. The mud was slippery, but the climb was an easy one. Soon I was standing on a dry mud floor near the top of the cave. I turned and looked across the river where the other three were waiting with the boats. A vine hung down over the cliff reaching almost to the water, and the arched cave opening was fringed with moss-covered stalactites. I lighted my headlamp and started into the inviting passage.

This could be a virgin cave, as I saw no footprints, although the floor was sticky mud. I continued beyond the sight of light and disturbed a small colony of bats. They fluttered at me, their squeaks barely audible. The passage paralleled the river for about two hundred feet and then narrowed to a mud fill. I retraced my steps and took a side passage away from the canyon, as this seemed to be the main corridor. The temperature of the caves in this part of Puerto Rico is in the low seventies, but the humidity is one hundred percent, making one feel quite warm. I stooped as I followed the passage, about six feet wide and not quite as high. There were several side corridors; and since the way seemed to be getting smaller, I checked out the side leads as I went along, but with no luck. Jack joined me; after some exploration, we determined that the cave did not seem to be much more extensive and we began to look for the forms of life so prevalent in tropical caves. We were rewarded in seeing several crabs, some crickets, and a *guavá*, the local name for a tailless whip scorpion. It is about the size of a silver dollar, with claw-like pincers, legs, and antennae that one could not cover with a dinner plate.

We left the cave, slid down the mud slope, and then washed ourselves in the river. When we were back again at the boats, we learned that Add had waded part way into the main cave. He came back to tell us that it was perfectly straight, and that he

Several hundred feet inside the first tunnel, looking back at the inverted keyhole entrance. *Roy Davis*.

could see the exit reflected in the flowing water at the other end. Don and Vito had taken compass sights; we all wondered how this phenomenon was formed. It was decided that a fault line probably provided this direct passage for the river.

We switched on our headlamps so that we could see the interior of the tunnel. As our eyes became accustomed to the dark, we saw that it was of almost uniform width (about twenty feet), of an undetermined depth, and of almost a uniform height—about fifteen feet. The water was calm and free of debris and rocks. It was an incredibly easy passage. We estimated the length to be between 700 and 800 feet from portal to portal, without a single indication of a side passage or twist in its whole length. We noted that the crevice running all along the ceiling was spotted with leaves, twigs, and pieces of bamboo, indicating that in high water this tunnel must flood completely, and provide a dam for the river. The hydrostatic head must flush out any blockage or debris that might be swept in.

We emerged into what seemed like a different river, as the water was deep and still. This section, with its high canyon walls, was a collapsed section of an ancient cave passage. Huge rounded blocks of limestone in the river made a channel for the water. This also made much better going for the boats, since the sluices were swift and deep and the portages much less frequent.

It was now about four o'clock, but the sun was well below the rim of the canyon. The sky was clouding up, and in a few minutes a few drops splattered on the water—then came a steady downpour. "I thought this was the dry season," mumbled Jack from the back of the boat.

We were soon drenched but plodded along, in and out of the boats, completely insensitive to our splashing and slipping on the limestone. We passed several more springs, which Vito and Don checked, and we were aware that there was indeed more water in the river now than when we had started. We were able to stay in the boats through several of the rapids and quickly covered the next quarter mile to the next known obstacle we had seen from the air.

This proved to be a spectacular sight, as the river surged against a solid rock wall, turned sharply left, and plunged through a magnificent natural arch. The arch was 60 feet wide, 100 feet high, and about 250 feet long. The top of the bridge was more than 150 feet thick, covered with green trees and vines that hung almost to the water. We beached the boats and gathered under a poncho to view the scene. The bridge was the last remnant of a cave, and festooned with twisted, moss-covered stalactites.

We moved inside the drip line out of the rain and began to check out both entrances. The walls were sheer except for a ledge that ran along one side about ten feet over the water level. "This

A natural bridge over the river shielded the first night's campsite. *Roy Davis*.

looks as if it would be the best spot to camp," Jack said, and he proceeded to wade across the river toward a ledge on the opposite side. This proved to be a fine idea; the ledge was above river level and was a shallow, dry cave high enough to stand in and broad and deep enough for all six of us to sleep in. The floor was covered with a thin layer of sand. The only deterrents were twigs and leaves stuck to the ceiling, indicating that at some time the cave completely flooded, but we decided that the river would have to rise at least ten feet to cause any concern. This did not appear to be likely, so we moved in, had supper, and spent an uneventful night.

When we opened our eyes at daylight, it was an inspiring sight to see a huge arch, one hundred feet overhead, covered with stalactites and illuminated from the reflected light off the river.

The camp began to stir, sleeping bags were stowed, bags packed, breakfast eaten, and the boats firmed up with a few pumps of the bellows. By eight o'clock we were again on the river.

We had gone less than a quarter of a mile when we saw a tributary coming in from the right. This was the Quebrada Pasto, one of the principal water sources for the Tanama. It had been dye-tested in the 1966 and 1970 groundwater tests made by Don Jordan. We tied up the boats and proceeded up the little trail along the river to see if there was a cave at the upstream end. The topographic map showed a little stream that disappeared underground for a distance of about 2,500 feet.

Vito and Don had brought machetes, which they proceeded to use as we followed the rather vague trail along the stream. The trail was narrow, wet, muddy, and slippery from the rain of the night before. We continued upstream for several thousand feet until we found several tributaries forming the headwaters. Following each of these we found more than one entrance or spring for the exit of the water.

One of them appeared to have passage, and Roy was soon on his way with me, following with compass and mapping pad. The entrance was about five feet wide and ten feet high, with about a foot of water on the floor. Just a few feet within the cave, the water deepened into a pool and the walls narrowed. As I debated with myself how to proceed, I heard Roy gasp as he plunged into the water and swam to the other side of a pool. He told me to come ahead so I proceeded, taking several bearings with the compass as I went. I found that it was possible to bridge the pool, and succeeded in passing the swimming area with only one wet leg. The passage twisted upward out of the water. This was probably the wet-weather exit of the water—the passage being useful for draining only in very heavy rains or flooding.

Roy had disappeared, so I proceeded mapping and climbing up the series of steps to a branch in the passage. As I sketched, I observed two whip scorpions two and a half inches long on the wall, quite unusual in the caves of Puerto Rico.

Roy returned in a few minutes and said that the cave continued but it required crawling, so we checked our measurements and returned to the entrance. The return to the boats was uneventful, except that I stopped to tie my shoe and kneeled down in a fire-ant nest. It was quite an unpleasant wade through the stream all the way back.

Another half mile and we were again in the canyon of the river which, with the addition of the side stream, was larger and deeper. The constriction of the canyon walls also added to the quantity of water that was now surging along. Rapids and falls became more frequent, and a few of the drops were so steep and high that we portaged the boats again. In the walls of the canyon on both sides we passed many small cave entrances. Finally, one was too large to pass by, and we paddled into the entrance and tied up the boats. This proved to be a large cave, again appearing to parallel the river. I sketched and mapped it, and Roy went another 1,500 feet only to appear on the river bank, as the cave was a side meander of the main river. There was ample evidence of flooding, and a large bat colony.

Our progress was excellent in the boats, and we recognized that it would not be long before we would reach the foot trail that descended from the radio telescope, where we had agreed to meet the support party.

In this section of the canyon the stream had cut large meanders and overhangs in the limestone. Some of these were deep enough to hide the stream completely from the air—another reason our air reconnaissance had not been successful. Many small springs poured out of the walls of the canyon and showered into the river. Most of them formed tufa or calcareous sinter as they cascaded down the wall. At times there would be springs entering from both sides, and at one place the sinter had built out

Huge limestone blocks fallen in the river from the canyon walls made it necessary to portage the boats. *Roy Davis*.

in a projecting ledge from either side of the river. The projections from each side were so great that they came within five feet of joining across the river in a natural bridge.

It was about four when we drew up all the boats on a gravel bar. Roy and Don found a small cave along the river to sleep in. The cave-camp was about fifty feet long and paralleled the river. It was open its entire length. Large stalactites had formed columns along the side which gave the cave a series of windows and doors overlooking the river.

Some of the springs that feed the Tanama River can be seen along the left bank. *Roy Davis*.

The campsite for the second night had stalactites of a natural limestone cave. *Roy Davis*.

Everyone managed to get out of wet clothes with the exception of me. Everything I had (except the sleeping bag) was completely soaked. Soon Add had the cooking water boiling and the food prepared. The world looked brighter (if not drier) to me.

It was now completely dark. We all had flashlights and carbide lamps, but the floor was so irregular that there was no place to sit or stand in a group. We had pulled two boats into the cave with us, and these were to be air mattresses for Vito and Don. This proved to be the best move of the day. The inverted boats, blocked up with seabags and bolsters, made two level platforms to sit on.

I began to get cold and uncomfortable. The only place I could find level or flat enough to put down my ground cloth and air mattress was at an oblique angle to the cave, with my feet sloping toward the river. Fortunately there was a stubby stalagmite that prevented the air mattress from sliding into the river, so I set up camp. My carbide lamp needed water, so I took off one sock and squeezed enough water to fill the lamp. I stripped down, wrung out my shorts, put them back on, and crawled into my sleeping bag. I looked at my watch and it was 6:30 P.M. "What a ridiculous hour to go to bed," I thought as I shone my light upward. The ceiling was a mass of stalactites only three feet over my head. I wondered which one was actively dripping, propped up on my poncho to act as a rain shield—and went to sleep.

About 11:00 P.M. I awoke and realized I had slid down about a foot toward the river, and that two feet away was a sheer drop to the water ten feet below. It brought to mind the time in the Navy when I had witnessed a burial at sea—the canvas-sewn body, the tilted plank, and the plunge into the ocean. The thought caused me to loosen the zipper on my sleeping bag, wiggle uphill in the bag, and stare into the darkness.

At dawn I was awakened by a groan from Vito. His rubber raft bed had not given him much back support. A glance out of the cave showed blue sky overhead, so we turned our thoughts to the day ahead. Add was up, and in a few minutes the hot coffee and his delicious pancakes set everything right for the day. We quickly broke camp, reluctantly putting on our wet, soggy gear, and cast off.

The decision was made that only one boat, unloaded, would attempt to go through the tunnel ahead. If this boat could make it, we would portage the gear over the mountain to the exit of the cave, and the other two boats would also travel unladen down the river.

Roy and Jack agreed to attempt to make the traverse through the cave. They agreed that they would go as far as they could—even down waterfalls where they could not get back—to the limit of the thirty-foot rope they had; this would permit them to be hauled back if necessary.

We transferred all the equipment to a gravel bar at the end of the observatory trail and piled it into one huge heap. Vito and Add set off up the trail to see a man who lived at the top of the hill, to arrange for horses to transport the gear up and over the mountain. They were to continue on over the hill to the exit ot the cave and there await the arrival of Roy and Jack.

Don and I remained on the river bank, enjoying the sun that appeared about 10:00 A.M. over the haystack hills. We dried some clothes, and then heard the clatter of horses coming down the trail. Vito had located the farmer and had sent him ahead for the gear.

Before we could separate the caving gear from the rescue gear, we heard a shout from up the trail. Jack and Roy and several others were heading toward us. The two had made the traverse with no trouble, and had even passed under the "impassable" sump at the end of the pool after exiting the cave. Watson Monroe, head of the support party, together with two men from the USGS and Barbara and Barry Beck, two spelunkers from the mainland who were living in Puerto Rico, arrived with them.

The obvious move was to go ahead with the other two boats. Roy and Jack and the others would go back up the mountain and choose a campsite for the night. Add, Vito, Don, and I would make the traverse and meet them at the campsite. All the gear was taken up the trail, leaving us the necessary lights and a pump for the boats.

We pushed off. The first little rapids were a smooth slip down several feet into a nearly still pool. As I turned back to check the other boat, I saw Vito seated on the pointed front, feet in the water, and Don in the rear with the paddle posed as a rudder. Their boat slowly approached the riffled water, picked up speed, and then, in a quick up-and-down motion, slid over the little drop like a flying carpet. We paddled ahead toward a barrier of rocks and toward an increasing roar of rushing water. I stood up in the boat and saw rapids ahead—then only green trees—no river. We pulled over to the side and waded ashore, pulling the boats along on a bow tether.

A fine jet of water was pouring through a thirty-inch crevice in the volcanic rocks. As it poured out, it arched through the air and dropped about twenty feet into a raging, foaming white flume for another fifty feet ahead. It was definitely not the way to go.

Portages were necessary to bypass some waterfalls. *Roy Davis.*

We skirted along the side of the river, down over a series of house-sized blocks; we pulled the boats over a little wooded peninsula and then down to another pool. The river had dropped about forty feet in less than one hundred feet. We put in the boats, paddled across the pool—and repeated the performance, for the river again leaped off through another crevice and plunged down over black rocks. We could only pick up the boats and trudge around the barrier.

After doing this three times, we were again back into limestone bedrock. The stream was still full of black volcanic boulders, but the gradient had softened; the rapids were longer; the pools shallower and the familiar canyon walls were closing in.

We knew we were approaching a cave even before we turned the next bend, as we could smell the acrid odor of bat guano. The view of the entrance was breathtaking: the river flowed into a huge corridor in the sheer cliff ahead. One hundred feet wide and a hundred feet high at the drip line, the corridor narrowed down later to a width of about sixty feet.

We stopped at the entrance to light our lamp. To our left was a vertical wall of limestone over 400 feet high; ahead was the huge opening, and to our right was dense vegetation obscuring a steeply rising hill. There was no reasonable way out but straight ahead. The river flowed just as actively into the entrance as it had behind us.

We got back into the boats and started into the cave. All was black ahead, and it was important to determine the headroom because our eyes had not yet become accustomed to the dark. It certainly must have been over a hundred feet. The river plunged over little falls near the entrance as it made a huge S turn. When we were 300 feet inside, there was still sufficient light from the entrance to light our way adequately. We shone our lights ahead and could see another pool just beyond a little rapid. Add and I decided we would try to run it in the boats to avoid clambering over slippery guano-covered rocks.

The sluice we decided to run was only a little wider than the boat and was flowing swiftly. There was sufficient water, but there was a rock in the middle that was invisible from the boat in the dim light.

We shot down the sluice, tobogganed up on the rock, and the boat started to jackknife. The stern began to fill with water as I was propelled over Add's shoulder into the river. Unfortunately I took Add with me, and we both floundered together. My hat, light, and paddle were swept away, but we found that we could get footing against the current and hold onto the boat. As soon as we realized neither of us was hurt, the situation almost seemed funny to us.

Vito and Don wisely decided to climb over the obstacle while we waded ashore, and as we were turning over the boat to empty out the water, Vito pointed out the reflection of light ahead. We were more than halfway through the cave and hardly out of sight of light. We had extra paddles along and so continued in the boats for an easy float toward what proved to be a beautiful, easy exit.

Although the trip through the cave was fairly easy, we had observed high water marks on the cave walls, and debris jammed in crevices as much as thirty feet above the river. This could be a treacherous trap in flood, and a few minutes later we could see why.

Ahead was the barrier Vito had seen two weeks before from the trail above. At this viewing from the river, it was a distinctive arched passage five feet high and ten to fifteen feet wide at the water line. It appeared to be a clear, straight passage through to the other side. Jack's and Roy's boat was standing up against one wall tied to the exposed roots of a tree.

Vito and Don stopped at the boat while Add and I paddled on to the cave opening in search of the paddle I had lost. We recovered it beneath a lacy waterfall which screened the entrance to the tunnel.

Closer examination of the tunnel revealed that it was actually a natural bridge formed of tufa, or calcareous sinter, a secondary formation like a cave formation, which actually spanned the river. The tiny stream that carried the dissolved calcite to form the bridge still flowed on its surface and dropped off each end into the river. Algae and vegetation in the pools above further upset the equilibrium of the spring water and accelerated the growth of the tufa. Beneath the bridge, stalactites covered the ceiling, but the basic rock had been transported by the water and had been redeposited in this unusual formation. The tunnel was about eight feet long and probably not more than twenty feet wide at the highest point. Jack and Roy had tried to climb over the bridge and found that it consisted of a series of limestone pools full of grasses and floating islands. It was actually a two-level stream system—one river above and one river below.

It is possible, although we have no way to prove it, that in extreme flooding the water of the main river might completely inundate this bridge, flush it clear below, and possibly scour out the pools above.

Add and I paddled back upstream to the others and tied up our boat along with the other two. We all climbed up a natural crevice in the side wall—an easy climb, using vines and roots—and in twenty minutes we came out in an open forty-five-degree slope. An almost level spot for a campsite was found further up the hill.

From this open vantage point we surveyed the region. We were on a steep slope. On top, on probably the only level spot around, was a two-story stilt house, the home of the farmer who scratched a living from this rugged land. He was able to grow coffee, bananas, oranges, and other fruits, but it must have been an endless job keeping the jungle from taking over his fields.

As we turned our backs on the house, we could see on the top of one of the highest haystack hills a tower that reached up another two hundred feet. This white masonry monolith was one of the three pylons of the ionospheric observatory. Two hundred astronomers, electronic specialists, and other scientists, probing the unknown in air-conditioned labs with sophisticated technical equipment, were less than a thousand feet away from our party—the first to explore the cave that runs literally beneath them.

The next morning we broke camp quickly after a very wet night. The sky looked as if it were going to rain all day, and our spirits were a bit low.

In two days we had traveled five river miles; it was five more miles to the bridge that spanned the river at Esperanza. At least three of those five miles had never been traveled. The air photos showed at least three more underground watercourses. One bright aspect was that no more volcanic rock was known to exist from here to the end of the river. This was encouraging, since we had found the river easy to navigate in the limestone regions.

We decided to leave all the camping and cooking gear, to take only the essentials with us, and try to make it in one day. The shore party was to meet us at 5:00 P.M. at the Esperanza bridge and we arranged for the farmer to bring the horses and move all the equipment up the hill.

We retraced our steps down the slope to the boats. The water level was unchanged, so we cast off and started through the first tunnel, which we had checked the night before.

We were again in unexplored territory. Soon there was no access to the river by land. The canyon was so narrow that in places the trees bridged the river. From the air there was no sign of the water for thousands of feet of its length. However, from the river it was incredibly beautiful and easy. The rapids were long and gradual, and the walls were sheer for ten to twenty feet on either side. We drifted along.

For hundreds of feet we could not even see the base rocks; water, seeping down from the heavy vegetation that clung to the steep canyon walls, covered the rocks. Streams cascaded down from the sides to form shower baths along the sides. In places, the vines and bushes drooped down almost to the water; and, on some of the sharp meanders of the river, the bed of the stream undercut the walls, leaving forty-foot projections over the water.

The next tunnel came into view, a natural bridge spanning the stream. The water was deep, the ceiling was ten to fifteen feet above the water, and the passage free and clear. We paddled and floated right through.

We were making remarkably good time, experiencing no obstacles or portages. We turned another bend and floated into another cave.

"That makes six," said Jack. "Three to the telescope and three this morning." Only one of these had been visible from the air.

Vito and Don stopped to take water samples of the springs,

and Don remarked that we had not seen any side caves. We decided that we might have been passing right by the tributaries which must be up above the river in the dense foliage. Possibly the little springs were coming from caves located above the river. Looking more carefully, we realized that there were many likely looking dark areas up in the trees, but we did not stop to investigate. We had more than two miles of unexplored river passage ahead that might present problems and delay our getting out of the canyon that day.

In a few minutes we came to another cave—number seven—and Jack and I could see as we came to the entrance that this would slow us down a little. Add and Roy were in the lead and were halfway through a hundred-foot tunnel. There was plenty of headroom and visibility to the exit ahead, but stuck in the middle of the cave was a wall-to-wall log jam of bamboo and floating debris. While it was only ten feet wide, there seemed no way to chimney or climb around it. We arrived in time to see Roy try to get out of the boat and stand on the bamboo. As soon as he put his weight on it, it began to sink and he flopped back in the boat after having sunk in the debris up to his waist. Although only about thirty feet long, the jam was too firm to let the boats pass through, but not solid enough to walk on.

Vito and Don were concerned about the bottom of their boat. They had already patched up one rip and did not think it would hold up if it had to be dragged over this jagged mass. Roy and Add meantime had discovered a means of sinking the debris in front by pulling the boat forward while standing on the flotsam ahead of it. They had several slips and spills, but were making progress. Don tried stepping out of his boat to urge it ahead and found that he was up to his neck in the mess and unable to pull himself up. Fortunately we were right behind him and were able to haul him aboard. Finally Add and Roy got to the head of the jam, freed the chock pole, and the jam began to break up.

We were floating free again and were able to paddle easily through the rest of the cave into open river. The water was deep, the shoals few, and the rapids full and clear. We were making good time and hoped to have a clear run to the bridge at Esperanza.

Soon another short, clear tunnel appeared that had not been visible on the air map. It was deep, with adequate headroom; we passed through it easily. The canyon now broadened into a valley and we knew we were approaching the last tunnel shown on the map. It was now possible (if not too practical) to climb out of the river from here, if necessary.

The final tunnel was several hundred feet long, but it had a skylight about three-quarters of the way through, permitting passage all the way without need of lights. The ceilings were forty feet high and the passage free and clear.

Just beyond this tunnel we saw our first evidence of civilization in the river when we floated over a stationary cable festooned with grass and debris. A farmer used it as a hand line to make a ford of the river. It was mute testimony to the difficulty of

The voyagers emerge from the last of the nine caves through which the river flows. *Roy Davis*.

crossing the river when the water was a little higher—one could easily be swept downstream.

A few more turns and we saw ahead the pilings and masonry structure of the bridge at Esperanza. In three days we had traversed ten miles of twisting river that had never before been navigated. It had gone underground nine times and dropped over six hundred feet in elevation. We had partially explored four caves, and the USGS men had taken water samples of fourteen springs along the way.

Another frontier had been opened up; as we beached the boats, we hoped many people would come after us to enjoy making this traverse of the Tanama.

Glossary

AA (pronounced "ah-ah") A rough, clinkery *lava* that does not flow smoothly, and thus does not form *lava caves.*

BELAY To secure a climber or rappeler by means of a safety rope; also the safety rope itself.

BIOSPELEOLOGY The scientific study of animal and plant life in *caves*. The specialist in this study is a biospeleologist.

BREAKDOWN Material which has fallen from the ceilings and walls of *caves,* in particular blocks of rock.

CALCITE A mineral, calcium carbonate, which is the main constituent of *limestone* and most *speleothems.*

CARBIDE A solid chemical, calcium carbide, which manufactures acetylene gas when exposed to water. Carbide is used as fuel for *carbide lamps.*

CARBIDE LAMP A miner's lamp used by many *cavers* that uses *carbide* for fuel. It burns with a bright yellow flame.

CAVE A natural underground orifice big enough for human beings to enter.

CAVER A person who explores or visits *caves,* especially a *spelunker.*

CAVERN Synonymous with *cave,* but sometimes elegantly used to indicate a large *cave.*

CHIMNEY A narrow, vertical crack or slot climbed by friction, by pressing arms, legs, or other parts of the body against opposite walls; also the act of climbing up or down a chimney.

COLUMN A *stalactite* and *stalagmite* that have grown together to form a continuous floor-to-ceiling *speleothem.*

COMMERCIAL CAVE A *cave* open to the public for a fee, usually with lights and improved walkways.

CONSERVATION The wise use of natural resources, in particular preserving them as much as possible under conditions that existed before the presence of man.

CRAWL or CRAWLWAY A *cave passage* so low that it can be traversed only on hands or knees, or the stomach.

DOME A vertical shaft rising from a horizontal *passage.* See also *domepit* and *pit.*

DOMEPIT or DOME-PIT A vertical shaft extending both above and below a horizontal passage. See also *dome* and *pit.*

ECOSYSTEM The interaction between living organisms and their environment.

FLOWSTONE A *speleothem* that coats the walls or floors and resembles a frozen waterfall.

FORMATION A *secondary* mineral deposit in a *cave.* The term *speleothem* is gradually replacing it in this context.

FOSSIL The remains or traces of any animal or plant that lived in the prehistoric past.

GLACIER CAVE A *cave* inside or at the bottom of a glacier.

GLACIÈRE A widely-used French synonym for *ice cave.*

GROTTO 1. A small alcove in a *cave.* 2. A chapter of the National Speleological Society.

GROUNDWATER Subsurface water which fills all the available cavities in the rock below the *water table.* The term is sometimes used to indicate all underground water, including that above the *water table.*

GUANO Excrement or dung of birds, bats, and other animals.

GYPSUM CAVE A *cave* developed in gypsum, or calcium sulfate.

GYPSUM FLOWER A picturesque gypsum *speleothem.*

HAYSTACK In tropical *limestone* regions, a conical hill which remains as a residue of *solution* and erosion.

HELICTITE (occasionally, helectite) A twisted, contorted *speleothem* that grows seemingly without regard to gravity.

HISTOPLASMOSIS A dust-borne disease, usually affecting the lungs, carried by bat or bird *guano.*

HYDROLOGY The study of the water in the earth.

HYDROLOGIST A scientist who studies the water in the earth.

ICE CAVE A cave in which ice can be found all year-long.

JOINT A crack or parting in a rock mass, often occurring in parallel groups or joint sets.

KARST A topography that often develops in *limestone* cave regions; it is characterized by *sinkholes,* disappearing creeks, few surface streams, and, of course, *caves.*

LAVA Molten rock extruded onto the earth's surface during volcanic activity.

LAVA CAVE or LAVA TUBE CAVE A *cave* formed in *lava* during volcanic activity.

LEAD (pronounced "leed") An unexplored small or narrow *passage.*

LIMESTONE A sedimentary rock composed primarily of the mineral *calcite.*

MARBLE A metamorphic rock, originally deposited as *limestone,* that has been recrystallized by heat and pressure.

PAHOEHOE (pronounced "pa-ho-ee-ho-ee") A smooth, ropy *lava* in which *lava caves* may develop.

PALEONTOLOGIST A scientist who specializes in the study of fossils and prehistoric life.

PASSAGE A general term for any underground channel.

PIT A vertical shaft descending from a horizontal *passage.* See also *dome* and *domepit.*

PLEISTOCENE The Ice Age; that period of time in earth history immediately preceding historic times when there were several episodes of glaciation. The term is sometimes used to include the present.

RAPPEL To descend a pit or cliff on a rope using various techniques.

SALTPETER (occasionally SALTPETRE) A mineral, potassium nitrate, a main ingredient of gunpowder, found in many *caves.*

SEA CAVE or LITTORAL CAVE A *cave* formed on the shore of a body of water through wave action.

SECONDARY DEPOSIT A feature or *speleothem* that develops after the *cave* itself has formed.

SINK 1. Short for *sinkhole.* 2. A stream or creek which disappears underground.

SINKHOLE A depression in the earth's surface caused by collapse of a portion of a *cave,* or by solution of the rock. Many cave entrances are in sinkholes.

SODASTRAW or SODASTRAW STALACTITE A thin-walled hollow stalactite that resembles a drinking straw.

SOLUTION In *speleology,* the chemical dissolving away of rock material.

SPELEAN Pertaining to *caves* or *speleology.*

SPELEOGENESIS Pertaining to the genesis or formation of *caves.*

SPELEOLOGIST A scientist who studies *caves* and related phenomena.

SPELEOLOGY The scientific study of *caves* and related phenomena.

SPELEOTHEM A *secondary* mineral deposit in a *cave.* Synonymous with *formation.*

SPELUNKER A person who explores *caves* primarily for the fun or sport of it. See *caver.*

STALACTITE A *speleothem* which hangs down from the ceiling. See *stalagmite.*

STALAGMITE A *speleothem* which grows up from the floor. See *stalactite.*

TROGLOBITE An animal, unable to survive on the surface, that must complete its life cycle in *caves.*

TROGLOPHILE An animal that can complete its life cycle either in *caves* or on the surface.

TROGLOXENE An animal that may habitually live in *caves,* but must periodically return to the surface to complete its life cycle.

VIRGIN CAVE A *cave* or part of a *cave* that has not been explored.

WATER TABLE A zone in the earth's surface below which all crevices and cavities are filled with water.

WALKING PASSAGE A *cave passage* large enough to permit one to walk upright.

For Further Reading

General Books on Caves and Caving

Most have additional information on specific caves and subjects mentioned in this volume, plus much other material.

Anderson, Jennifer, *Cave Exploring*. Association Press, New York, 1974.

Folsom, Franklin, *Exploring American Caves*. Collier, New York, 1962.

Halliday, William R., *American Caves and Caving*. Harper and Row, New York, 1974.

Halliday, William R., *Depths of the Earth*, Revised edition. Harper and Row, New York, 1977.

Hovey, Horace C., *Celebrated American Caverns*. Johnson Reprint, New York, 1970. Reprint of the 1896 cave classic.

McClurg, David R., *The Amateur's Guide to Caves and Caving*. Stackpole Books, Harrisburg (Pennsylvania), 1973.

Mohr, Charles E., and Howard N. Sloane, *Celebrated American Caves*. Rutgers University Press, New Brunswick (New Jersey), 1955.

Waltham, Tony, *Caves*. Crown, New York, 1975.

Cave Life

Barbour, Roger W., and Wayne H. Davis, *Bats of America*. University Press of Kentucky, Lexington, 1969.

Mohr, Charles E., and Thomas L. Poulson, *The Life of the Cave*. McGraw-Hill, New York, 1966.

Mammoth Cave

Brucker, Roger W., and Richard A. Watson, *The Longest Cave*. Alfred A. Knopf, New York, 1976.

Lawrence, Joe, Jr., and Roger W. Brucker, *The Caves Beyond*. Zephyrus Press, Teaneck (New Jersey), 1975. Reprint of the 1955 edition.

Meloy, Harold, *Mummies of Mammoth Cave.* Micron Publishing Co., Shelbyville (Indiana), 1968.

Watson, Patty Jo, *Archaeology of the Mammoth Cave Area.* Academic Press, New York, 1974.

Other Caves

These books are about caves mentioned in this volume, or caves of the immediate geographic area.

Conn, Herb and Jan, *The Jewel Cave Adventure.* Zephyrus Press, Teaneck (New Jersey), 1977.

Gurnee, Russell and Jeanne, *Discovery at the Rio Camuy.* Crown, New York, 1974.

Jackson, George F., *The Story of Wyandotte Cave.* Speleobooks, Albuquerque (New Mexico), 1975.

Index

Page numbers in **boldface** refer to illustrations and maps.